Math Mammoth
Fractions 1

By Maria Miller

Contents

Introduction

Math Mammoth Fractions 1 is the first book of two that cover all aspects of fraction arithmetic. This book covers the concepts of fraction and mixed number, equivalent fractions, adding and subtracting like and unlike fractions, adding and subtracting mixed numbers, comparing fractions, and measuring in inches. The book *Math Mammoth Fractions 2* covers the rest of fraction arithmetic: simplifying fractions and multiplication and division of fractions.

I have made a set of videos to match many of the lessons in this book. You can access them at https://www.mathmammoth.com/videos/fractions_1

Studying fractions involves lots of rules, and many students learn them only mechanically, not really understanding the underlying concepts and principles. Students then end up making lots of mistakes because they confuse the different rules, and either apply the wrong one or apply the right rule but don't remember it quite right. All this can make students even fear fractions.

To avoid that, we use the visual model of a pie divided into slices all the way through the book. It is a natural model because a circle can be divided into any number of circle sectors (slices). When students work with this model from lesson to lesson, they will eventually start seeing these pies in their mind. This, in turn, gives them the ability to do many of the easier fraction calculations mentally. It also enables students to really UNDERSTAND these concepts, and not just learn mechanical rules.

You are welcome to use manipulatives along with the book; however the visual pie model is probably sufficient for most students in the fifth grade level. I have also included (in the appendix) printable cut-outs for fractions from halves to twelfths.

The first lesson, *Fraction Terminology*, explains various fraction terminology such as proper and improper fractions, like and unlike fractions, and so on. The student can refer back to this information as needed.

The lesson *Review: Mixed Numbers* briefly goes over how to write fractions as mixed numbers and vice versa. The concepts in this lesson are prerequisites for the rest of the book.

The next lessons cover adding and subtracting mixed numbers when the fractional parts have the same denominator. The lessons are well illustrated with "pies" to help the student visualize how regrouping works in the context of fractions.

Then, it is time to study equivalent fractions, as a prerequisite for adding unlike fractions. Equivalent fractions are presented as parts that have been split further. The rule is to multiply both the numerator and the denominator by the same number, but try to emphasize the terminology of "splitting the existing parts into so-and-so many pieces" or something similar. That should help students to understand the concept instead of memorizing a mechanical rule.

Adding and Subtracting Unlike Fractions is an introductory lesson in the sense that the student is not yet introduced to the rule for finding the common denominator. Instead, in this lesson, the common denominator is either given, or the student figures it out using pictures.

Finding the (Least) Common Denominator emphasizes the idea that we need to find a common denominator, and then convert the fractions to like fractions before adding.

Next, students add and subtract mixed numbers with unlike fractional parts and solve some word problems relating to the concept.

Then we cover the concept of comparing fractions. The lesson presents several strategies, many of which have been studied in earlier grades. New for this level is to convert the fractions to equivalent, like fractions, just like when adding fractions.

In the last two lessons of the book, students learn to measure items using 16th parts of an inch and to make line plots from measurement data.

Answers are appended.

I wish you success in teaching math!
Maria Miller, the author

Helpful Resources on the Internet

We have compiled a list of external Internet resources that match the topics in this book. This list of links includes web pages that offer:

- **online practice** for concepts;
- online **games**, or occasionally, printable games;
- **animations** and interactive **illustrations** of math concepts;
- **articles** that teach a math concept.

We heartily recommend you take a look at the list. Many of our customers love using these resources to supplement the bookwork. You can use the resources as you see fit for extra practice, to illustrate a concept better, and even just for some fun. Enjoy!

https://l.mathmammoth.com/blue/fractions1

SCAN ME

Fraction Terminology

As we study fraction operations, it is important that you understand the terms, or words, that we use. This page is for reference. You can post it on your wall or even make your own fraction poster based on it. Some of the terms below you already know; some we will study in this book.

$\dfrac{3}{11}$ The top number is the **numerator**. It *enumerates,* or numbers (counts), *how many* pieces there are.

The bottom number is the **denominator**. It *denominates,* or names, *what kind* of parts they are.

A mixed number has two parts: a whole-number part and a fractional part.

For example, in $2\dfrac{3}{7}$, the whole-number part is 2, and the fractional part is $\dfrac{3}{7}$.

The mixed number $2\dfrac{3}{7}$ actually means $2 + \dfrac{3}{7}$.

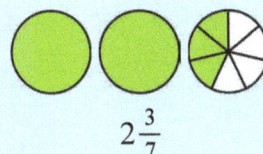

$2\dfrac{3}{7}$

Like fractions have the same denominator. They have the same kind of parts. It is easy to add and subtract like fractions, because all you have to do is look at *how many* of that kind of part there are.

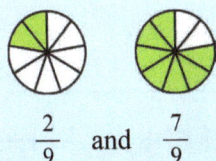

$\dfrac{2}{9}$ and $\dfrac{7}{9}$ are like fractions.

Unlike fractions have a different denominator. They have different kinds of parts. It is a little more complicated to add and subtract unlike fractions. You need to first change them into like fractions. Then you can add or subtract them.

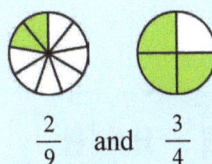

$\dfrac{2}{9}$ and $\dfrac{3}{4}$ are unlike fractions.

A proper fraction is a fraction that is less than 1 (less than a whole pie). 2/9 is a proper fraction.

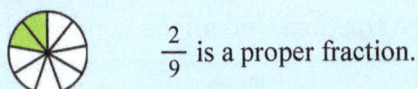

An improper fraction is more than 1 (more than a whole pie). Being a *fraction*, it is written as a fraction and *not* as a mixed number.

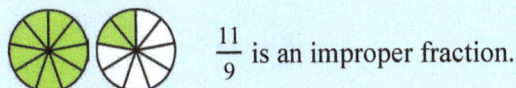

$\dfrac{2}{9}$ is a proper fraction.

$\dfrac{11}{9}$ is an improper fraction.

Equivalent fractions are equal in value. If you think in terms of pies, they have the same amount of "pie to eat," but they are written using different denominators, or are "cut into different kinds of slices."

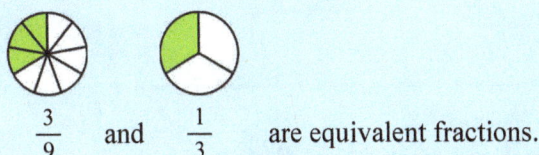

$\dfrac{3}{9}$ and $\dfrac{1}{3}$ are equivalent fractions.

Simplifying or reducing a fraction means that, for a given fraction, you find an equivalent fraction that has a "simpler," or smaller, numerator and denominator. (It has fewer but bigger slices.)

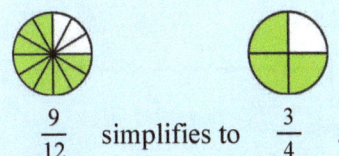

$\dfrac{9}{12}$ simplifies to $\dfrac{3}{4}$.

Review: Mixed Numbers

1. Write the mixed numbers that these pictures illustrate.

a.

b.

c.

2. Draw pictures that illustrate these mixed numbers.

a. $3\frac{2}{6}$

b. $4\frac{7}{8}$

3. Write the mixed number that is illustrated by each number line.

a.

```
0    1    2    3
```

b.

```
0    1    2    3
```

4. Write the fractions and mixed numbers that the arrows indicate.

```
    a.   b.              c.   d.
0    1    2    3    4    5
```

5. Mark the fractions on the number line. $\frac{9}{8}$, $\frac{22}{8}$, $\frac{13}{8}$, $\frac{24}{8}$, $\frac{11}{8}$

```
0              1              2              3
```

Example 1. To write $4\frac{2}{9}$ as a fraction, we *count* all the ninths:

- Each pie has nine ninths, so the four complete pies have 4 × ____ = ____ ninths.
- Additionally, the incomplete pie has ____ ninths.
- The total is ____ ninths or $\dfrac{}{}$.

6. Write as mixed numbers and as fractions.

a. $1\dfrac{2}{5} = \dfrac{}{5}$

b.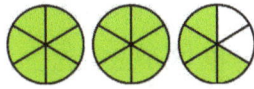

c.

d.

e.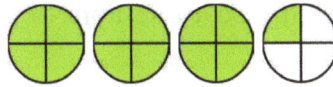

Shortcut: $5\dfrac{3}{4} = \dfrac{23}{4}$
×

Numerator: 5 × 4 + 3 = 23

Denominator: 4

Multiply the whole number times the denominator, then add the numerator, to get the number of fourths, or the numerator for the fraction. The denominator will remain the same.

7. Explain how the shortcut works, and why. Use the image on the right as an example.

$5\dfrac{9}{13}$
× +

8. Write as fractions. Think of the shortcut.

a. $7\dfrac{1}{2}$

b. $6\dfrac{2}{3}$

c. $8\dfrac{3}{9}$

d. $6\dfrac{6}{10}$

e. $2\dfrac{5}{11}$

f. $8\dfrac{1}{12}$

g. $2\dfrac{5}{16}$

h. $4\dfrac{7}{8}$

Fractions to mixed numbers

Example 2. To write a fraction, such as 58/7, as a mixed number, we need to figure out:

- how many *whole* "pies" there are, and
- how many *individual slices* are left over.

In the case of 58/7, each whole "pie" will have 7 sevenths. (How do you know?) So we ask:

- How many 7s are there in 58? (Those make the whole pies!)
- After the 7s are gone, how many "slices" are left over?

Finish this example in the next exercise.

9. Refer to the example above. How many 7s are there in 58? _____

   After that, how many "slices" are left over? _____

   What math operation helps you with the above?

   Now, use the answers to the above questions to write 58/7 as a mixed number.

10. Rewrite the "division problems with remainders" as "fractions changed to mixed numbers."

a. $47 \div 4 = 11$ R3 $\dfrac{47}{4} = 11\dfrac{3}{4}$	**b.** $35 \div 8 = 4$ R3 $\underline{} = \square \ \underline{}$	**c.** $19 \div 2 = \underline{}$ R $\underline{}$ $\underline{} = \square \ \underline{}$
d. $35 \div 6 = \underline{}$ R $\underline{}$ $\underline{} = \square \ \underline{}$	**e.** $72 \div 10 = \underline{}$ R $\underline{}$ $\underline{} = \square \ \underline{}$	**f.** $22 \div 7 = \underline{}$ R $\underline{}$ $\underline{} = \square \ \underline{}$

The Shortcut: Think of the fraction bar as a *division* symbol, and DIVIDE. The quotient tells you the whole number part, and the remainder tells you the numerator of the fractional part.

11. Write these fractions as mixed numbers (or as whole numbers, if you can).

a. $\dfrac{62}{8} =$	**b.** $\dfrac{16}{3} =$	**c.** $\dfrac{27}{5} =$	**d.** $\dfrac{32}{9} =$
e. $\dfrac{7}{2} =$	**f.** $\dfrac{25}{4} =$	**g.** $\dfrac{50}{6} =$	**h.** $\dfrac{32}{5} =$
i. $\dfrac{24}{11} =$	**j.** $\dfrac{39}{3} =$	**k.** $\dfrac{57}{8} =$	**l.** $\dfrac{87}{9} =$

Adding Mixed Numbers

Review: adding like fractions

Example 1. Here, 5/9 and 7/9 are *like* fractions: they have the same denominator (same kind of parts). To add them, simply add the numerators.

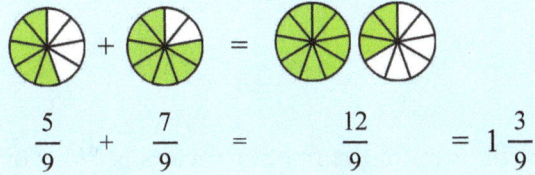

Why does the denominator not change?

Because the *kind* of parts they are is not changing.

$$\frac{5}{9} + \frac{7}{9} = \frac{12}{9} = 1\frac{3}{9}$$

The answer, 12/9, is an improper fraction (more than one whole), so we write the final answer as a mixed number.

---

1. The calculation on the right shows a common student error. What is the error?

   Fix the calculation.

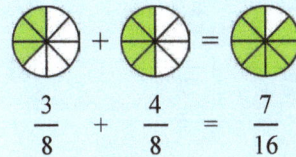

   $$\frac{3}{8} + \frac{4}{8} = \frac{7}{16}$$

2. Add. If your final answer is more than one, write it as a mixed number.

a. $\frac{5}{10} + \frac{3}{10} =$	**b.** $\frac{7}{8} + \frac{5}{8} =$	**c.** $\frac{7}{12} + \frac{5}{12} =$
d. $\frac{1}{4} + \frac{3}{4} + \frac{3}{4} =$	**e.** $\frac{4}{5} + \frac{3}{5} + \frac{2}{5} =$	

In this lesson we only deal with mixed numbers that have *like* fractional parts (the same denominator.) To add them, simply **add the whole number parts and the fractional parts separately:**

$$1\frac{1}{7} + 5\frac{3}{7} = 6\frac{4}{7}$$

You can also add in columns →

$$\begin{array}{r} 2\frac{5}{8} \\ + 3\frac{2}{8} \\ \hline 5\frac{7}{8} \end{array}$$

---

3. Add.

a. $3\frac{2}{4} + 7\frac{1}{4} =$	**c.** $\begin{array}{r} 2\frac{1}{4} \\ + 11\frac{2}{4} \\ \hline \end{array}$	**d.** $\begin{array}{r} 28\frac{6}{12} \\ + 4\frac{4}{12} \\ \hline \end{array}$
b. $15\frac{3}{9} + 3\frac{5}{9} =$		

Often the sum of the fractional parts is more than one whole pie. Look at this example <u>carefully</u>:

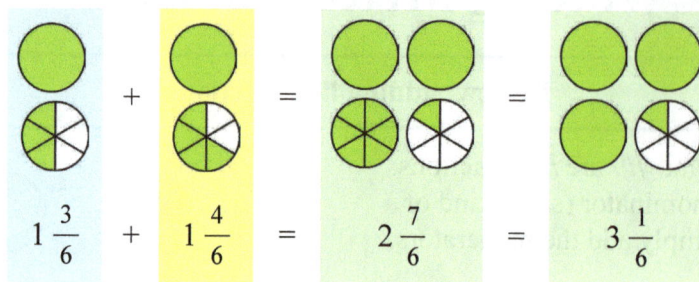

$$1\frac{3}{6} \quad + \quad 1\frac{4}{6} \quad = \quad 2\frac{7}{6} \quad = \quad 3\frac{1}{6}$$

Here, the sum of the fractional parts is 7/6. Think of that as 1 1/6, and <u>add the one whole to the sum of the whole numbers</u> (which was 2) to get 3. The final answer is 3 1/6.

4. These mixed numbers have a fractional part that is more than one "pie." Write them in such a way that the fractional part is less than one. The first one is done for you. You can use manipulatives or draw fraction pictures to help.

| a. $3\frac{3}{2} = 4\frac{1}{2}$ | b. $1\frac{11}{9} =$ | c. $6\frac{7}{4} =$ | d. $3\frac{13}{8} =$ |

5. Write the addition sentences that the pictures illustrate and then add.

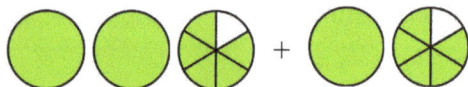

a.

b.

6. Add.

| a. $3\frac{2}{3} + 8\frac{1}{3} =$ | b. $4\frac{4}{5} + 1\frac{3}{5} =$ | c. $6\frac{8}{9} + 1\frac{2}{9} =$ |

7. The sides of a triangle measure 7 3/8 inches, 5 7/8 inches, and 3 4/8 inches. What is its perimeter?

8. Add.

| a. $4\frac{3}{7}$ $+\ 5\frac{5}{7}$ ____ $9\frac{8}{7} = 10\frac{1}{7}$ | b. $3\frac{3}{5}$ $+\ 3\frac{4}{5}$ ____ $=$ | c. $4\frac{6}{9}$ $+\ 2\frac{7}{9}$ ____ | d. $7\frac{6}{8}$ $+\ 2\frac{7}{8}$ ____ |

9. Find the missing addend. You can draw more pie pictures to help.

a. $1\frac{1}{4}$ + ___ = 5	**b.** $2\frac{2}{3}$ + ___ = 7

10. Find the missing addends. You can draw pictures to help.

a. $2\frac{1}{4}$ + $1\frac{1}{4}$ + ___ = 5	**b.** $3\frac{2}{5}$ + $2\frac{2}{5}$ + ___ = 8

11. Change the final answer so that the fractional part (13/6) is less than one. How many additional whole pies can you make out of it?

$$1\frac{5}{6} \ + \ 1\frac{3}{6} \ + \ 1\frac{5}{6} \ = \ 3\frac{13}{6} \ = \ \boxed{\ } \ \frac{\quad}{\quad}$$

12. Add.

a. $3\frac{1}{6}$ + $2\frac{5}{6}$ =	**b.** $4\frac{4}{5}$ + $1\frac{2}{5}$ + $5\frac{2}{5}$ =
c. $6\frac{4}{8}$ + $1\frac{6}{8}$ + $1\frac{7}{8}$ =	**d.** $3\frac{6}{10}$ + $3\frac{8}{10}$ + $\frac{9}{10}$ =

13. Jeremy runs 2 ¼ miles four days a week.
    Robert runs 3 ½ miles three times a week.
    Which boy runs more in one week?
    How much more?

14. Add the mixed numbers.

a. $10\frac{7}{9}$ $2\frac{5}{9}$ + $3\frac{8}{9}$ ___ =	**b.** $1\frac{5}{11}$ $3\frac{9}{11}$ + $2\frac{8}{11}$ ___ =	**c.** $2\frac{5}{6}$ $5\frac{4}{6}$ + $2\frac{3}{6}$ ___ =	**d.** $1\frac{7}{10}$ $\frac{9}{10}$ + $10\frac{6}{10}$ ___ =

Subtracting Mixed Numbers 1

Strategy 1: Renaming / regrouping. In this method we first cut one of the whole pies into slices, and join these slices with the existing slices. After that, we can subtract.

Example 1.

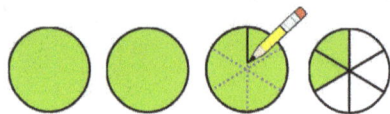

$$3\frac{2}{6} - 1\frac{5}{6}$$
$$\downarrow$$
$$2\frac{6}{6} + \frac{2}{6} - 1\frac{5}{6}$$
$$\downarrow$$
$$2\frac{8}{6} - 1\frac{5}{6} = 1\frac{3}{6}$$

At first we have three uncut pies and 2/6 more. We cut one of the whole pies into sixths. We end up with only two whole (uncut) pies and 8 sixths.

We say that 3 2/6 has been **renamed** as 2 8/6. Now we can subtract 1 5/6 easily.

We can also solve this problem by writing the mixed numbers one under the other.

$$\begin{array}{r} 2 \\ \cancel{3}\,\cancel{\frac{2}{6}}^{\,\frac{8}{6}} \\ -\,1\frac{5}{6} \\ \hline 1\frac{3}{6} \end{array}$$

We **regroup** (borrow) 1 whole pie as 6 sixths. There are already 2 sixths in the fractional parts column, so we add the 6/6 and 2/6 and write 8/6 in place of 2/6. Now we can subtract the 5/6.

Example 2.

$$2\frac{1}{8} - \frac{5}{8}$$
$$\downarrow$$
$$1\frac{9}{8} - \frac{5}{8} = 1\frac{4}{8}$$

Or:

We regroup 1 whole as 8/8. The 8/8 and the existing 1/8 make a total of 9/8.

$$\begin{array}{r} 1 \\ \cancel{2}\,\cancel{\frac{1}{8}}^{\,\frac{9}{8}} \\ -\,\frac{5}{8} \\ \hline 1\frac{4}{8} \end{array}$$

1. Don't subtract anything. Divide ONE whole pie into fractional parts and rename the mixed number.

a. $2\frac{1}{6}$ is renamed as

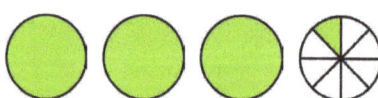

b. $3\frac{1}{8}$ is renamed as

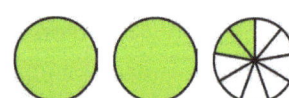

c. $2\frac{2}{9}$ is renamed as

d. $2\frac{3}{5}$ is renamed as

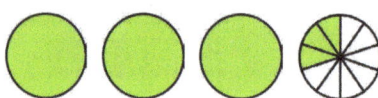

e. $3\frac{3}{10}$ is renamed as

f. $2\frac{1}{4}$ is renamed as

14

2. Rename, then subtract. Be careful. Use the pie pictures to check your calculation.

a. $4\frac{2}{9} - 1\frac{8}{9}$

$= 3\frac{11}{9} - 1\frac{8}{9} =$

b. $5\frac{3}{12} - 2\frac{7}{12}$

$= 4\frac{}{12} - 2\frac{7}{12} =$

c. $5\frac{7}{10} - 3\frac{9}{10}$

$= \frac{}{} - 3\frac{9}{10} =$

d. $4\frac{3}{8} - 1\frac{7}{8}$

$= \frac{}{} - 1\frac{7}{8} =$

3. Regroup (if necessary) and subtract.

a.
$$2\frac{}{9}$$
$$\cancel{3\frac{4}{9}}$$
$$- \frac{8}{9}$$
_____

b.
$$7\frac{4}{9}$$
$$- 2\frac{7}{9}$$
_____

c.
$$12\frac{9}{12}$$
$$- 6\frac{11}{12}$$
_____

d.
$$8\frac{3}{14}$$
$$- 5\frac{9}{14}$$
_____

e.
$$14\frac{7}{9}$$
$$- 3\frac{5}{9}$$
_____

f.
$$11\frac{5}{21}$$
$$- 7\frac{15}{21}$$
_____

g.
$$26\frac{4}{19}$$
$$- 14\frac{15}{19}$$
_____

h.
$$10\frac{3}{20}$$
$$- 5\frac{7}{20}$$
_____

Strategy 2: Subtract in Parts. First, subtract what you can from the fractional part of the minuend. Then subtract the rest from one of the whole pies. Study the examples.

Example 3. $2\frac{1}{8} - \frac{5}{8}$

$$= 2\frac{1}{8} - \frac{1}{8} - \frac{4}{8}$$

$$= \quad 2 \quad - \frac{4}{8}$$

$$= \quad 1\frac{8}{8} \quad - \frac{4}{8} = 1\frac{4}{8}$$

First we take away only 1/8, which leaves 2 whole pies. Then we subtract the rest (4/8) from one of the whole pies.

Example 4. $3\frac{2}{9} - 2\frac{7}{9}$

$$= 3\frac{2}{9} - 2\frac{2}{9} - \frac{5}{9}$$

$$= \quad 1 \quad - \frac{5}{9}$$

$$= \quad \frac{9}{9} \quad - \frac{5}{9} = \frac{4}{9}$$

We cannot subtract 7/9 from 2/9. So, first we subtract 2 and 2/9, which leaves 1 whole pie. The rest, 5/9, is subtracted from the last whole pie.

4. Subtract in parts. Remember: you can *add* to check a subtraction problem.

a. $2\frac{2}{6} - \frac{5}{6}$

b. $3\frac{1}{5} - 2\frac{3}{5}$

c. $3\frac{2}{7} - 2\frac{6}{7} =$

5. Ellie had 4 yards of material. She needed ⅞ yard for making a skirt, and she made two. How much material is left?

6. Subtract in two parts. Write a subtraction sentence.

a. Cross out $1\frac{5}{9}$.

b. Cross out $1\frac{11}{12}$.

Example 5. Look at Mia's math work: $7\frac{1}{6} - 2\frac{5}{6} = 9\frac{6}{6} = 10$. Can you see why it is wrong?

If you have 7-and-a-fraction, and you subtract 2-and-a-fraction, you cannot get 10 as an answer! Nor could you get 2-and-a-fraction or 3-and-a-fraction. The answer to this problem should be either 5-and-a-fraction or 4-and-a-fraction.

In reality, Mia was *adding* instead of subtracting. (If you have ever done that, you are not alone—it is a common error.)

Always check if your answer is reasonable (not too small, nor too big).

7. Find the answers that are not reasonable, and redo those problems.

a. $8\frac{1}{5} - 3\frac{3}{5} = 4\frac{3}{5}$	**b.** $4\frac{2}{8} - 1\frac{7}{8} = 1\frac{1}{8}$	**c.** $12\frac{4}{13} - 9\frac{8}{13} = 21\frac{12}{13}$
d. $11\frac{2}{15} - 6\frac{6}{15} = 2\frac{11}{15}$	**e.** $7\frac{1}{20} - 3\frac{7}{20} = 3\frac{14}{20}$	**f.** $6\frac{14}{100} - 2\frac{29}{100} = 5\frac{85}{100}$

8. Two sides of a triangle measure 3 ⅝ in, and the perimeter of the triangle is 10 ⅛ in. How long is the third side of the triangle?

9. A recipe for chocolate cookies calls for 1 ¾ cups of flour and Harry is making a <u>double</u> batch. However, he *only* has ¾ cup of flour! How much more flour would Harry need to have enough to make the cookies?

Puzzle Corner   If you repeatedly subtract $1\frac{5}{9}$, starting from 14, will you eventually reach zero?

Subtracting Mixed Numbers 2
(This lesson is optional.)

> **Strategy 3 (optional):** Subtract the whole numbers and the fractional parts separately. If you get a *negative* fraction, treat the "minus" as a subtraction symbol.

> **Example 1.** $6\frac{2}{10} - 2\frac{5}{10} = ?$
>
> - Subtract the whole numbers: $6 - 2 =$ **4** ;
>
> - Subtract the fractions: $\frac{2}{10} - \frac{5}{10} = -\frac{3}{10}$.
>
> - Combine the two results: $\mathbf{4} - \frac{3}{10} = 3\frac{7}{10}$.

> **Example 2.** $8\frac{1}{7} - 5\frac{6}{7} = ?$
>
> - Subtract the whole numbers: $8 - 5 =$ **3** ;
>
> - Subtract the fractions: $\frac{1}{7} - \frac{6}{7} = -\frac{5}{7}$.
>
> - Combine the two results: $\mathbf{3} - \frac{5}{7} = 2\frac{2}{7}$.

1. Subtract using any strategy. Remember to check that your answer is reasonable. If you subtract 5-and-something − 1-and-something, your answer cannot be 2-and-something.

a. $5\frac{3}{8} - 1\frac{7}{8} =$	**b.** $9\frac{2}{15} - 5\frac{8}{15} =$
c. $7\frac{11}{30} - 4\frac{9}{30} =$	**d.** $16\frac{5}{12} - 4\frac{11}{12} =$

2. You have 3 ¾ kg of ground beef. Your neighbor buys ¾ kg of it and you use ¾ kg to make meatballs. How much beef do you have left?

3. Subtract.

a. $\begin{array}{r} 5\frac{1}{11} \\ -\ 3\frac{2}{11} \\ \hline \end{array}$	**b.** $\begin{array}{r} 6\frac{6}{7} \\ -\ 1\frac{5}{7} \\ \hline \end{array}$	**c.** $\begin{array}{r} 6\frac{2}{15} \\ -\ 1\frac{9}{15} \\ \hline \end{array}$

4. Find the missing minuend or subtrahend.

a. $\quad - \ 2\frac{1}{5} = 3\frac{2}{5}$	**b.** $\quad - \ 2\frac{6}{12} = 3\frac{5}{12}$	**c.** $\ 7\frac{8}{9} - \quad = 4\frac{1}{9}$

5. Add and subtract.

a. $\ 2\frac{1}{4} + 5\frac{3}{4} - 3\frac{2}{4} =$	**b.** $\ 4\frac{5}{6} + 6\frac{3}{6} - 1\frac{4}{6} =$
c. $\ 9\frac{3}{8} + 2\frac{7}{8} - 3\frac{6}{8} =$	**d.** $\ 7\frac{7}{12} + 3\frac{11}{12} - 1\frac{2}{12} =$

6. Here are some more subtraction problems for additional practice, if your teacher so indicates. Color the answer squares as indicated. Use extra paper for calculations.

a. (yellow) $\qquad 5\frac{2}{9} - 2\frac{7}{9}$

b. (blue) $\qquad 7\frac{8}{15} - 4\frac{11}{15}$

c. (blue) $\qquad 5\frac{6}{11} - 3\frac{2}{11}$

d. (yellow) $\qquad 4\frac{1}{9} - 2\frac{3}{9}$

e. (green) $\qquad 17\frac{2}{9} - 4\frac{5}{9}$

f. (green) $\qquad 5\frac{1}{11} - 3\frac{9}{11}$

g. (green) $\qquad 10\frac{1}{12} - 4\frac{7}{12}$

h. (yellow) $\qquad 4\frac{3}{10} - 2\frac{3}{10}$

i. (green) $\qquad 5\frac{1}{10} - 3\frac{9}{10}$

j. (yellow) $\qquad 8\frac{1}{8} - 2\frac{5}{8}$

k. (blue) $\qquad 7\frac{1}{11} - 3\frac{5}{11}$

l. (blue) $\qquad 9\frac{7}{8} - 3\frac{1}{8}$

m. (yellow) $\qquad 15\frac{3}{12} - 10\frac{4}{12}$

$4\frac{11}{12}$	$1\frac{2}{10}$	$12\frac{6}{9}$	$2\frac{4}{9}$
$3\frac{7}{11}$			$2\frac{4}{11}$
$2\frac{12}{15}$		2	$6\frac{6}{8}$
$1\frac{7}{9}$	$1\frac{3}{11}$	$5\frac{6}{12}$	$5\frac{4}{8}$

19

Equivalent Fractions 1

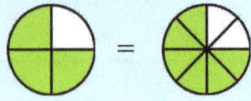

These two fractions are **equivalent fractions** because they picture the same, or equal, amounts. You could say that you get to "eat" the same amount of "pie" either way.

In the second picture, **each slice** has been **split or cut into two pieces**. The arrows show into how many new pieces each piece was split.

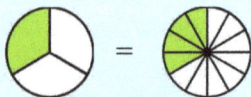

Each slice has been split into four.

BEFORE: 1 colored piece, 3 total.
AFTER: 4 colored pieces, 12 total.

Notice that we get *four* times as many colored pieces and *four* times as many total pieces. This means that both the numerator and the denominator get multiplied by 4.

When all of the pieces are split the same way, both the number of colored pieces (the numerator) and the total number of pieces (the denominator) get multiplied by the same number.

1. Connect the pictures that show equivalent fractions. Write the name of each fraction beside its picture.

2. Make a chain of equivalent fractions. Notice the patterns!

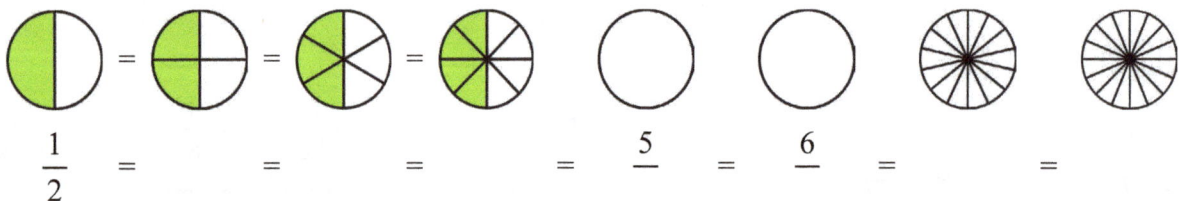

3. Split the pieces by drawing the new pieces in the right-hand picture. Write the equivalent fractions.

a. Split each piece <u>in two</u>.	b. Split each piece <u>into three</u>.	c. Split each piece <u>in two</u>.
×2 $$\frac{2}{5} = \frac{}{}$$ ×2	×3 $$\frac{1}{2} = \frac{}{}$$ ×3	×2 $$\frac{2}{3} = \frac{}{}$$ ×2
d. Split each piece <u>in two</u>.	e. Split each piece <u>into three</u>.	f. Split each piece <u>in two</u>.
=	=	=
g. Split each piece <u>in two</u>.	h. Split each piece <u>in two</u>.	i. Split each piece <u>into five</u>.
=	=	

4. Write the equivalent fraction. Imagine or draw the helping arrows.

a. Split each piece into four.	b. Split each piece in two.	c. Split each piece into six.	d. Split each piece into four.	e. Split each piece into five.
$\frac{3}{4} = \frac{}{}$	$\frac{5}{8} = \frac{}{}$	$\frac{1}{2} = \frac{}{}$	$\frac{2}{7} = \frac{}{}$	$\frac{1}{4} = \frac{}{}$
f. Split each piece into three.	g. Split each piece into ten.	h. Split each piece into eight.	i. Split each piece into seven.	j. Split each piece into eight.
$\frac{2}{7} = \frac{}{}$	$\frac{5}{8} = \frac{}{}$	$\frac{1}{2} = \frac{}{}$	$\frac{3}{5} = \frac{}{}$	$\frac{3}{7} = \frac{}{}$

5. Figure out how many ways the pieces were split and write the missing numerator or denominator.

a. Pieces were split into <u>three</u>.	**b.** Pieces were split into ____ .	**c.** Pieces were split into ____ .	**d.** Pieces were split into ____ .	**e.** Pieces were split into ____ .
$\times 3$ $\dfrac{4}{7} = \dfrac{}{21}$ $\times 3$	$\times \square$ $\dfrac{4}{5} = \dfrac{}{20}$ $\times \square$	$\times \square$ $\dfrac{1}{6} = \dfrac{}{18}$ $\times \square$	$\times \square$ $\dfrac{6}{7} = \dfrac{}{14}$ $\times \square$	$\times \square$ $\dfrac{2}{3} = \dfrac{8}{}$ $\times \square$
f. $\dfrac{3}{10} = \dfrac{9}{}$	**g.** $\dfrac{2}{11} = \dfrac{6}{}$	**h.** $\dfrac{4}{7} = \dfrac{}{56}$	**i.** $\dfrac{1}{6} = \dfrac{}{54}$	**j.** $\dfrac{7}{8} = \dfrac{}{64}$

6. Mark the equivalent fractions on the number lines.

a. $\dfrac{2}{3} = \dfrac{\square}{12} = \dfrac{\square}{24}$

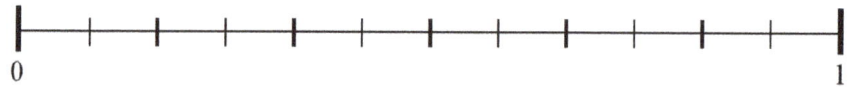

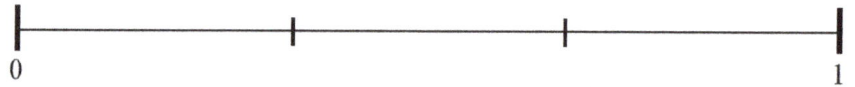

b. $\dfrac{5}{6} = \dfrac{\square}{12} = \dfrac{\square}{24}$

c. Find and mark two fractions on the 12th parts number line that do *not* have an equivalent fraction on the 3rd parts number line. Write them here →

d. Find and mark two fractions on the 24th parts number line that do *not* have an equivalent fraction on the 12th parts number line. Write them here →

7. A family of four baked two pizzas that were the same size, each cut into 12 slices. Dad ate 1/2 of one pizza, Mom and Cindy ate 1/3 of a pizza each, and Derek ate the rest.

a. Write the given fractions in the table (Dad, Mom, and Cindy). Then for each, write an equivalent fraction, using 1/12 parts.

b. Figure out what part of a pizza Derek ate.

	Fraction	Equivalent fraction
Dad	$\dfrac{1}{2}$	
Mom		
Cindy		
Derek		

Equivalent Fractions 2

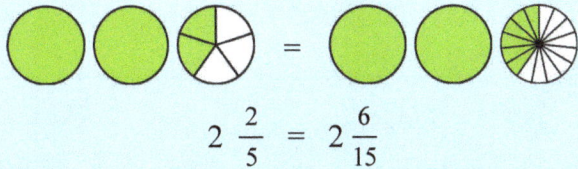

$$2\frac{2}{5} = 2\frac{6}{15}$$

Here you see the mixed number 2 2/5 changed into an equivalent mixed number 2 6/15. Actually, we only changed the fractional part, 2/5, into the equivalent fraction 6/15. The whole-number part did not change.

These pictures show the fraction 7/3 converted into an equivalent fraction 14/6. Now, 7/3 is a fraction, not a mixed number. You can see that from the picture because the whole pies have been split into fractional pieces. We consider it as seven thirds (slices).

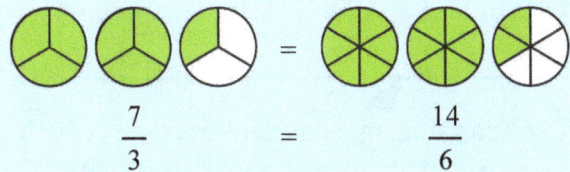

$$\frac{7}{3} = \frac{14}{6}$$

Also, 7/3 is an **improper fraction** because its value is 1 or more. (Of course, 14/6 is also.) A **proper fraction** is a fraction whose value is less than 1.

We use equivalent fractions also with mixed numbers and with improper fractions.

1. These are improper fractions. Split the slices in the right-hand picture. Write the equivalent fraction.

a. Split each slice into three.

$$\frac{7}{4} = \frac{}{}$$

b. Split each piece in two.

$$\frac{12}{5} = \frac{}{}$$

c. Split each piece in two.

$$\frac{5}{3} = \frac{}{}$$

d. Split each slice into four.

$$\frac{4}{2} = \frac{}{}$$

2. Fill in what is missing.

a. $5\frac{7}{10} = \boxed{}\frac{\boxed{}}{80}$	**b.** $5\frac{7}{10} = \boxed{}\frac{28}{\boxed{}}$	**c.** $\frac{9}{4} = \frac{63}{\boxed{}}$	**d.** $\frac{7}{1} = \frac{\boxed{}}{6}$
e. $6\frac{2}{9} = \boxed{}\frac{12}{\boxed{}}$	**f.** $2\frac{3}{4} = \boxed{}\frac{\boxed{}}{28}$	**g.** $\frac{16}{5} = \frac{32}{\boxed{}}$	**h.** $\frac{8}{3} = \frac{\boxed{}}{15}$

3. Here, we have written the number 3 as a *fraction*, as 3/1. Write the number 3 as a fraction using other kinds of parts, also.

whole pies	halves	thirds	fourths	fifths	tenths	hundredths
$\dfrac{3}{1}$	$\dfrac{\ }{2}$					

4. Write the number 2 1/2 as a fraction using...

halves	fourths	sixths	eighths	tenths	twentieths	hundredths
$\dfrac{\ }{2}$						

5. If you can find an equivalent fraction, write it. If you cannot, cross out the whole problem.

a. $\dfrac{5}{7} = \dfrac{\ }{28}$ The pieces were split into _____ .	**b.** $\dfrac{2}{5} = \dfrac{\ }{18}$ The pieces were split into _____ .	**c.** $\dfrac{1}{4} = \dfrac{\ }{14}$ The pieces were split into _____ .	**d.** $\dfrac{2}{3} = \dfrac{\ }{12}$ The pieces were split into _____ .	**e.** $\dfrac{5}{6} = \dfrac{8}{\ }$ The pieces were split into _____ .
f. $\dfrac{1}{6} = \dfrac{\ }{28}$ The pieces were split into _____ .	**g.** $\dfrac{2}{9} = \dfrac{\ }{63}$ The pieces were split into _____ .	**h.** $\dfrac{5}{4} = \dfrac{\ }{32}$ The pieces were split into _____ .	**i.** $\dfrac{1}{3} = \dfrac{5}{\ }$ The pieces were split into _____ .	**j.** $\dfrac{3}{8} = \dfrac{8}{\ }$ The pieces were split into _____ .

6. Explain in your own words when a problem of equivalent fractions is *not possible* to do. Use an example problem or problems in your explanation.

7. Make chains of equivalent fractions. Pay attention to the *patterns* in the numerators and in the denominators.

a. $\dfrac{3}{4} = \dfrac{\ }{8} = \dfrac{\ }{12} = \quad = \quad = \quad = \quad = \quad$

b. $\dfrac{5}{3} = \dfrac{10}{6} = \dfrac{\ }{9} = \quad = \quad = \quad = \quad = \quad$

Adding and Subtracting Unlike Fractions

Cover the page below the line. Then try to figure out the addition problems below.
(Parent/teacher: You can also give this to your student(s) on a separate paper, to explore on their own.)

$\frac{1}{3}$ + $\frac{1}{2}$ = What fraction would this be??

$\frac{1}{3}$ + $\frac{1}{4}$ = What fraction would this be??

$\frac{1}{3}$ + $\frac{1}{2}$

↓ ↓

$\frac{2}{6}$ + $\frac{3}{6}$ = $\frac{5}{6}$

$\frac{1}{3}$ + $\frac{1}{4}$

↓ ↓

$\frac{4}{12}$ + $\frac{3}{12}$ = $\frac{7}{12}$

Did you find a solution to the problems above?

It is this:

We convert the fractions so that they become *like* **fractions** (with a same denominator), using **equivalent fractions**.

Then we can add (or subtract).

1. Write the fractions shown by the pie images. Convert them into equivalent fractions with the same denominator (like fractions), and then add them. Color the missing parts.

a.
$\frac{1}{2}$ + $\frac{1}{4}$

↓ ↓

___ + ___ = ___

b.

___ + ___

↓ ↓

___ + ___ = ___

c.

___ + ___

↓ ↓

___ + ___ = ___

2. Convert the fractions to like fractions first, then add or subtract. In the bottom problems (d-f), you need to figure out what kind of pieces to use, but the *top* problems (a-c) will help you do that!

a.
$$\frac{1}{2} + \frac{1}{6}$$
$$\downarrow \qquad \downarrow$$
$$+ \quad = $$
$$\frac{}{} + \frac{1}{6} = \frac{}{}$$

b.
$$\frac{1}{8} + \frac{1}{4}$$
$$\downarrow \qquad \downarrow$$
$$+ \quad = $$
$$\frac{1}{8} + \frac{}{} = \frac{}{}$$

c.
$$\frac{1}{6} + \frac{1}{4}$$
$$\downarrow \qquad \downarrow$$
$$+ \quad = $$
$$\frac{}{} + \frac{}{} = \frac{}{}$$

d.
$$\frac{5}{6} - \frac{1}{2}$$
$$\downarrow \qquad \downarrow$$
$$\frac{5}{6} - \frac{}{} = \frac{}{}$$

e.
$$\frac{5}{8} - \frac{1}{4}$$
$$\downarrow \qquad \downarrow$$
$$\frac{}{} - \frac{}{} = \frac{}{}$$

f.
$$\frac{5}{6} - \frac{1}{4}$$
$$\downarrow \qquad \downarrow$$
$$\frac{}{} - \frac{}{} = \frac{}{}$$

3. Convert the fractions to like fractions first, then add or subtract. In the bottom problems (d-f), you need to figure out what kind of pieces to use, but the *top* problems (a-c) will help you do that!

a.
$$\frac{1}{2} + \frac{1}{8}$$
$$\downarrow \qquad \downarrow$$
$$+ \quad = $$
$$\frac{}{} + \frac{}{} = \frac{}{}$$

b.
$$\frac{3}{10} + \frac{1}{5}$$
$$\downarrow \qquad \downarrow$$
$$+ \quad = $$
$$\frac{}{} + \frac{}{} = \frac{}{}$$

c.
$$\frac{2}{5} + \frac{1}{2}$$
$$\downarrow \qquad \downarrow$$
$$+ \quad = $$
$$\frac{}{} + \frac{}{} = \frac{}{}$$

d.
$$\frac{1}{2} + \frac{3}{8}$$
$$\downarrow \qquad \downarrow$$
$$\frac{}{} + \frac{}{} = \frac{}{}$$

e.
$$\frac{9}{10} - \frac{2}{5}$$
$$\downarrow \qquad \downarrow$$
$$\frac{}{} - \frac{}{} = \frac{}{}$$

f.
$$\frac{4}{5} - \frac{1}{2}$$
$$\downarrow \qquad \downarrow$$
$$\frac{}{} - \frac{}{} = \frac{}{}$$

4. Use vertical and/or horizontal lines to cut the *first* fraction so that both fractions will have the same kind of parts. Then add.

a. ____ + $\dfrac{3}{4}$ =

b. $\dfrac{}{8}$ + $\dfrac{5}{8}$ =

c. ____ + $\dfrac{5}{6}$ =

Now do cuts in *both* fractions:

d. $\dfrac{}{10}$ + $\dfrac{}{10}$ =

e. $\dfrac{}{}$ + $\dfrac{}{}$ =

f. $\dfrac{}{}$ + $\dfrac{}{}$ =

5. **a.** Fill in the table based on the problems above. What kind of parts did the two fractions have at first? What kind of parts did you have in the end?

Types of parts:	Converted to:	Types of parts:	Converted to:
2nd parts and 8th parts	_8th_ parts	2nd parts and 5th parts	_____ parts
2nd parts and 4th parts	_____ parts	3rd parts and 5th parts	_____ parts
3rd parts and 6th parts	_____ parts	3rd parts and 2nd parts	_____ parts

b. Now think: How can you know into what kind of parts to convert the fractions that you are adding? Can you see any patterns or rules in the table above?

6. Challenge: If you think you know what kind of parts to convert these fractions into, then try these problems. Do not worry if you don't know how to do them—we will study this in the next lesson.

a. $\dfrac{1}{2}$ + $\dfrac{2}{3}$

↓   ↓

$\dfrac{}{}$ + $\dfrac{}{}$ =

b. $\dfrac{2}{3}$ − $\dfrac{2}{5}$

↓   ↓

$\dfrac{}{}$ − $\dfrac{}{}$ =

c. $\dfrac{1}{3}$ + $\dfrac{3}{4}$

↓   ↓

$\dfrac{}{}$ + $\dfrac{}{}$ =

Finding the (Least) Common Denominator

Before adding or subtracting unlike fractions, first convert them into *like* fractions.

Before the conversion, we need to decide into what kinds of parts to convert them—in other words, what will be the new denominator for both. We call this denominator **the common denominator**, because all of the converted fractions will have this same denominator in common.

To do the actual conversion, use the principles you have learned concerning equivalent fractions.

Example 1.

$$\frac{1}{6} + \frac{5}{9}$$

$$\downarrow \qquad \downarrow$$

$$\frac{3}{18} + \frac{10}{18} = \frac{13}{18}$$

Notice that we used 18 as the common denominator. Why 18?

You will find out soon, on the next page.

For now, notice that 1/6 is converted into 3/18 and 5/9 is converted into 10/18 using the rule for writing equivalent fractions. See the sidebar on the right for a reminder. →

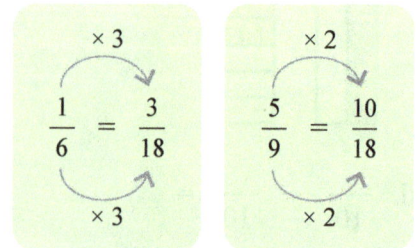

$$\times 3 \qquad\qquad \times 2$$
$$\frac{1}{6} = \frac{3}{18} \qquad\qquad \frac{5}{9} = \frac{10}{18}$$
$$\times 3 \qquad\qquad \times 2$$

1. You are given the common denominator. Convert the fractions using the rule for equivalent fractions. Then add or subtract. Note: sometimes you need to convert only one fraction, not both.

a. $\dfrac{1}{3} + \dfrac{3}{5}$ $\downarrow \qquad \downarrow$ $\dfrac{}{15} + \dfrac{}{15} =$	**b.** $\dfrac{6}{7} - \dfrac{1}{2}$ $\downarrow \qquad \downarrow$ $\dfrac{}{14} - \dfrac{}{14} =$	**c.** $\dfrac{1}{6} + \dfrac{2}{5}$ $\downarrow \qquad \downarrow$ $\dfrac{}{30} + \dfrac{}{30} =$
d. $\dfrac{5}{9} - \dfrac{1}{3}$ $\downarrow \qquad \downarrow$ $\dfrac{}{9} - \dfrac{}{9} =$	**e.** $\dfrac{1}{8} + \dfrac{3}{4}$ $\downarrow \qquad \downarrow$ $\dfrac{}{8} + \dfrac{}{8} =$	**f.** $\dfrac{5}{7} - \dfrac{2}{3}$ $\downarrow \qquad \downarrow$ $\dfrac{}{21} - \dfrac{}{21} =$
g. $\dfrac{2}{5} + \dfrac{1}{4}$ $\downarrow \qquad \downarrow$ $\dfrac{}{20} + \dfrac{}{20} =$	**h.** $\dfrac{5}{6} - \dfrac{3}{4}$ $\downarrow \qquad \downarrow$ $\dfrac{}{12} - \dfrac{}{12} =$	**i.** $\dfrac{3}{4} - \dfrac{3}{7}$ $\downarrow \qquad \downarrow$ $\dfrac{}{28} - \dfrac{}{28} =$

The common denominator has to be a <u>multiple</u> of each of the denominators.

This means that the common denominator has to be in the multiplication table of the individual denominators. In other words, the individual denominators have to "go into" the common denominator, or the common denominator has to be divisible by the individual denominators.

Example 2.

$$\frac{2}{3} + \frac{1}{5} = \frac{}{15} + \frac{}{15}$$

The common denominator must be a multiple of 5 and also a multiple of 3. Fifteen will work (it is in the multiplication table of 5 and also of 3).

Example 3.

$$\frac{3}{8} - \frac{1}{6} = \frac{}{24} - \frac{}{24}$$

Check the multiples of 8 (the skip-counting list): 0, 8, 16, 24, *etc.*
Compare to the multiples of 6: 0, 6, 12, 18, 24, 30, *etc.*
We notice that **24** is the smallest number that is in both lists.

Example 4.

$$\frac{7}{8} + \frac{3}{4} = \frac{7}{8} + \frac{}{8}$$

We need a number that 4 can "go into" and that 8 can "go into." Actually, the smallest such number is 8. So in this case, the 7/8 does not need to be converted; you just convert 3/4 into 6/8.

2. Find a common denominator (c.d.) that will work with these fractions.

fractions to add/subtract	c.d.
a. 4th parts and 5th parts	
b. 3rd parts and 7th parts	
c. 10th parts and 2nd parts	

fractions to add/subtract	c.d.
d. 4th parts and 12th parts	
e. 2nd parts and 7th parts	
f. 9th parts and 6th parts	

3. Let's add and subtract. Use the common denominators you found above.

a. $\dfrac{4}{5} + \dfrac{1}{4}$
↓   ↓
$\dfrac{}{20} + \dfrac{}{20} =$

b. $\dfrac{2}{3} - \dfrac{1}{7}$
↓   ↓
$\dfrac{}{} - \dfrac{}{} =$

c. $\dfrac{3}{10} + \dfrac{1}{2}$
↓   ↓
$\dfrac{}{} + \dfrac{}{} =$

d. $\dfrac{4}{12} + \dfrac{1}{4}$
↓   ↓
$\dfrac{}{} + \dfrac{}{} =$

e. $\dfrac{1}{2} - \dfrac{2}{7}$
↓   ↓
$\dfrac{}{} - \dfrac{}{} =$

f. $\dfrac{5}{6} - \dfrac{4}{9}$
↓   ↓
$\dfrac{}{} - \dfrac{}{} =$

You can always multiply the denominators to get a common denominator. However, you can often find a *smaller number* than the one you get by multiplying the denominators.

$\dfrac{7}{10}$ and $\dfrac{1}{15}$	We could use $10 \times 15 = 150$, but let's look at the lists of multiples: Multiples of 15: 0, 15, **30**, 45, 60, 75 ... Multiples of 10: 0, 10, 20, **30**, 40, 50, ... So, 30 works as well, and is smaller! It is the **least common denominator**.
$\dfrac{2}{7}$ and $\dfrac{1}{6}$	One possibility is $7 \times 6 = 42$, but let's check the multiples of 6 to make sure: Multiples of 6: 0, 6, 12, 18, 24, 30, 36, 42, 48, ... None of those are in the multiplication table of 7, except 42. So, 42 is the **L**east **C**ommon **D**enominator (LCD).

4. Find the least common denominator (LCD) for adding or subtracting these fractions. You may use the space for writing out lists of multiples.

fractions	LCD
a. $\dfrac{5}{12} + \dfrac{3}{8}$ $\downarrow$ $\qquad$ $\downarrow$ $+$ $\qquad$ $=$	
b. $\dfrac{7}{4} - \dfrac{9}{11}$ $\downarrow$ $\qquad$ $\downarrow$ $-$ $\qquad$ $=$	
c. $\dfrac{1}{12} + \dfrac{1}{9}$ $\downarrow$ $\qquad$ $\downarrow$ $+$ $\qquad$ $=$	
d. $\dfrac{7}{8} - \dfrac{4}{9}$ $\downarrow$ $\qquad$ $\downarrow$ $-$ $\qquad$ $=$	

Add and Subtract: More Practice

Example 1. Amy added: $\frac{2}{5} + \frac{2}{7} = \frac{4}{35}$, but that is not right. How can you tell?

Because the answer, 4/35, is a very small number, close to zero. It is much less than 2/5!

Example 2. Sam added: $\frac{5}{6} + \frac{3}{4} = \frac{8}{10}$, but it is wrong. How can you tell?

The answer, 8/10, is less than 1, whereas both 5/6 and 3/4 are more than 1/2, so their sum is more than 1.

Always check if your answer is reasonable (not too big, not too small).

Once you master adding and subtracting unlike fractions, you will have learned the *hardest part* of fraction math! **Congratulations!** Multiplication and division of fractions are actually easier.

1. Think whether each fraction in these calculations is close to 0, close to 1/2 or close to 1. Mark the answers that are unreasonable. You don't have to find the right answers.

$$\frac{1}{10} + \frac{8}{9} = \frac{20}{45} \qquad \frac{7}{12} + \frac{1}{18} = \frac{23}{36} \qquad \frac{7}{10} + \frac{8}{9} = \frac{143}{90} \qquad \frac{5}{8} + \frac{9}{16} = \frac{14}{32}$$

$$\frac{2}{11} + \frac{1}{10} = \frac{91}{110} \qquad \frac{5}{12} - \frac{2}{5} = \frac{43}{60} \qquad \frac{4}{5} - \frac{1}{13} = \frac{47}{65} \qquad \frac{8}{15} + \frac{5}{6} = \frac{12}{3}$$

2. Which of the possible answers is reasonable?

a. $\frac{2}{13} + \frac{15}{16}$

- ☐ $\frac{17}{208}$
- ☐ $\frac{129}{208}$
- ☐ $\frac{227}{208}$

b. $\frac{7}{8} - \frac{1}{9}$

- ☐ $\frac{55}{72}$
- ☐ $\frac{79}{72}$
- ☐ $\frac{12}{72}$

c. $\frac{5}{12} - \frac{1}{8}$

- ☐ $\frac{79}{96}$
- ☐ $\frac{28}{96}$
- ☐ $\frac{56}{96}$

3. Explain why these answers are wrong. You don't have to find the correct answer.

a. Amanda added: $\frac{5}{8} + \frac{5}{8} = \frac{10}{16}$

b. Robert subtracted: $\frac{7}{9} - \frac{1}{2} = \frac{6}{7}$

4. Explain why Olivia's answer must be wrong. Then find the correct answer.

Olivia subtracted: $\dfrac{7}{12} - \dfrac{1}{4} = \dfrac{1}{2}$ Correct answer:

How can you tell it is wrong?

5. Write the letters that match the answers in the boxes to solve the riddle. Check that your answers are reasonable.

Why did the banana go to the doctor?

| $\dfrac{23}{36}$ | $\dfrac{5}{6}$ | $\dfrac{7}{15}$ | $\dfrac{3}{10}$ | $\dfrac{1}{10}$ | $\dfrac{4}{15}$ | $\dfrac{5}{6}$ | | $\dfrac{23}{30}$ | $\dfrac{7}{45}$ | | $\dfrac{8}{15}$ | $\dfrac{31}{30}$ | $\dfrac{17}{28}$ |

☐ ☐ ☐ ☐ ☐ ☐ ☐   ☐ ☐   ☐ ☐ ☐

| $\dfrac{17}{21}$ | $\dfrac{9}{10}$ | $\dfrac{9}{35}$ | | $\dfrac{83}{72}$ | $\dfrac{11}{24}$ | $\dfrac{11}{24}$ | $\dfrac{27}{40}$ | $\dfrac{23}{30}$ | $\dfrac{5}{9}$ | $\dfrac{1}{6}$ | | $\dfrac{1}{14}$ | $\dfrac{104}{63}$ | $\dfrac{7}{6}$ | $\dfrac{13}{20}$ |

☐ ☐ ☐   ☐ ☐ ☐ ☐ ☐ ☐ ☐   ☐ ☐ ☐ ☐ !

L $\dfrac{1}{2} + \dfrac{2}{3}$ $\dfrac{\ }{6} + \dfrac{\ }{6} =$	**W** $\dfrac{1}{5} + \dfrac{1}{3}$	**S** $\dfrac{2}{5} - \dfrac{2}{15}$
E $\dfrac{2}{6} + \dfrac{1}{2}$	**C** $\dfrac{2}{3} - \dfrac{1}{5}$	**G** $\dfrac{5}{6} - \dfrac{2}{3}$
A $\dfrac{1}{10} + \dfrac{1}{5}$	**I** $\dfrac{1}{3} + \dfrac{13}{30}$	**U** $\dfrac{3}{5} - \dfrac{1}{2}$

These problems also pertain to the riddle on the previous page.

E $\frac{1}{3} + \frac{1}{8}$	**T** $\frac{5}{9} - \frac{2}{5}$	**N** $\frac{2}{3} + \frac{1}{7}$
E $\frac{5}{6} - \frac{3}{8}$	**O** $\frac{1}{2} + \frac{4}{10}$	**W** $\frac{4}{7} - \frac{1}{2}$
L $\frac{4}{5} - \frac{3}{20}$	**S** $\frac{6}{7} - \frac{1}{4}$	**B** $\frac{2}{9} + \frac{5}{12}$
T $\frac{6}{7} - \frac{3}{5}$	**A** $\frac{1}{5} + \frac{5}{6}$	**P** $\frac{11}{8} - \frac{2}{9}$
N $\frac{2}{3} - \frac{1}{9}$	**E** $\frac{10}{7} + \frac{2}{9}$	**L** $\frac{7}{8} - \frac{1}{5}$

Puzzle Corner

Find the fractions that can go into the puzzles.

Hint: If the answer has a denominator of 15, think what the denominators of the two fractions could have been.

	+		= $\frac{13}{15}$
+		+	
	+		= $\frac{5}{12}$
=		=	
$\frac{5}{6}$		$\frac{9}{20}$	

	+		= $\frac{13}{42}$
+		+	
	+		= $\frac{17}{72}$
=		=	
$\frac{7}{24}$		$\frac{16}{63}$	

33

Adding and Subtracting Mixed Numbers

In this lesson, we will be adding and subtracting **mixed numbers with unlike fractional parts.** Here's how:

1. First convert the unlike fractional parts into like fractions.

2. Then add or subtract the mixed numbers.

Example 1.

$$2\frac{1}{2} \Rightarrow 2\frac{4}{8}$$

$$+ 1\frac{7}{8} \Rightarrow + 1\frac{7}{8}$$

$$3\frac{11}{8} \Rightarrow 4\frac{3}{8}$$

Notice that the answer, 3 11/8, has a fractional part that is more than one (an improper fraction). Therefore, we need to write it as 4 3/8.

---

1. First convert the fractional parts into like fractions, then add.

a. $6\frac{2}{3} \Rightarrow 6\frac{}{15}$

$+ 3\frac{1}{5} \Rightarrow + 3\frac{}{}$

b. $10\frac{1}{8} \Rightarrow$

$+ 3\frac{2}{5} \Rightarrow$

c. $17\frac{1}{16} \Rightarrow$

$+ 3\frac{3}{8} \Rightarrow$

---

2. First convert the fractional parts into like fractions, then add. Lastly, change your final answer so that the fractional part is not an improper fraction.

a. $4\frac{1}{2} \Rightarrow 4\frac{}{10}$

$+ 3\frac{4}{5} \Rightarrow 3\frac{}{10}$

$\Rightarrow$

b. $5\frac{5}{6} \Rightarrow$

$+ 7\frac{2}{3} \Rightarrow$

$\Rightarrow$

c. $3\frac{5}{6} \Rightarrow$

$+ 2\frac{7}{8} \Rightarrow +$

$\Rightarrow$

d. $9\frac{5}{7} \Rightarrow$

$+ 7\frac{2}{3} \Rightarrow$

$\Rightarrow$

Example 2. Study how we can write the same problem and its solution either horizontally or vertically.

Horizontally:

$$2\frac{1}{2} - 1\frac{2}{3} = 2\frac{3}{6} - 1\frac{4}{6}$$

$$\downarrow$$

$$= 1\frac{9}{6} - 1\frac{4}{6} = \frac{5}{6}$$

Notice how 2 3/6 is **renamed** as 1 9/6. This is the same process as regrouping in the vertical solution.

Vertically:

$$2\frac{1}{2} \Rightarrow 2\frac{3}{6} \Rightarrow 1\frac{9}{6}$$
$$-1\frac{2}{3} \qquad\qquad -1\frac{4}{6}$$
$$\overline{\qquad\qquad\qquad \frac{5}{6}}$$

3. Solve. You can use the pies to help.

a. $2\frac{3}{4} - 1\frac{3}{8}$

b. $3\frac{1}{2} - 1\frac{1}{3}$

c. $3\frac{1}{3} - 1\frac{4}{9}$

4. First convert the fractional parts into like fractions, then subtract. You may need to regroup.

a. $5\frac{1}{2} \Rightarrow$

$-2\frac{4}{5} \Rightarrow$

b. $15\frac{4}{8} \Rightarrow$

$-8\frac{5}{6} \Rightarrow$

c. $16\frac{5}{9} \Rightarrow$

$-10\frac{1}{2} \Rightarrow$

d. $4\frac{1}{6} \Rightarrow$

$-2\frac{3}{5} \Rightarrow$

e. $11\frac{1}{12} \Rightarrow$

$-3\frac{1}{4} \Rightarrow$

f. $8\frac{2}{9} \Rightarrow$

$-2\frac{3}{4} \Rightarrow$

5. Spot the unreasonable answers, and correct them.

a. ½ kg of meat and another ¼ kg of meat makes 2/6 kg of meat.	b. Mike: "As of today, ⅕ of the job is done, and tomorrow I'll do half of it. That means 6/5 of it will be done."
c. ⅜ cups of flour and another ½ cup of flour will make ⅞ cups of flour.	d. Mia: "Last week I jogged 9 ½ km, and this week 7 ¾ km. So, last week I jogged 1 ¾ km more than this week."

6. Sally needs 1 1/4 meters of material to make a blouse and 8/10 of a meter to make a skirt.

   a. Find how many meters of material she needs for both of them.

   b. Now use *decimals* to solve the same problem. Which way do you feel is easier?

7. Henry's two heaviest school books weigh 1 3/4 lb and 1 11/16 lb.

   a. What is their total weight in *pounds*?

   b. Remember that 1 lb = 16 oz. Now change the total weight into pounds and ounces.

Comparing Fractions

1. What strategy could you use to find out which of the two fractions is greater?
   Do not simply state which fraction is greater, but explain the strategy. You may also work with a partner. For any pair of fractions that you cannot find a strategy, no worries — you will learn that in the following pages.

   a. $\dfrac{5}{12}$ and $\dfrac{3}{12}$

   b. $\dfrac{4}{7}$ and $\dfrac{4}{5}$

   c. $\dfrac{1}{2}$ and $\dfrac{6}{11}$

   d. $\dfrac{7}{6}$ and $\dfrac{8}{9}$

   e. $\dfrac{11}{20}$ and $\dfrac{7}{15}$

   f. $\dfrac{13}{15}$ and $\dfrac{7}{8}$

Sometimes it is easy to know which fraction is the greater of the two. Study the examples!

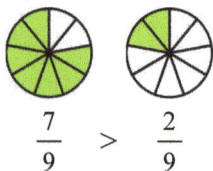

$$\frac{7}{9} > \frac{2}{9}$$

With **like fractions**, all you need to do is to check **which fraction has more "slices,"** and that fraction is greater.

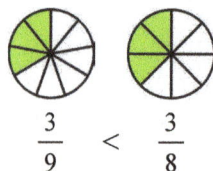

$$\frac{3}{9} < \frac{3}{8}$$

If both fractions have the **same number of pieces**, then the one with bigger pieces is greater.

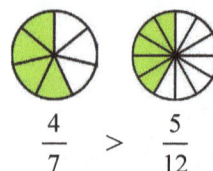

$$\frac{4}{7} > \frac{5}{12}$$

Sometimes you can **compare to 1/2**. Here, 4/7 is clearly more than 1/2, and 5/12 is clearly less than 1/2.

2. Find the fractions that are more than 1/2.

$$\frac{2}{5} \qquad \frac{5}{8} \qquad \frac{5}{6} \qquad \frac{5}{12} \qquad \frac{10}{21} \qquad \frac{8}{14} \qquad \frac{4}{10} \qquad \frac{28}{50}$$

3. Compare the fractions, and write > , < , or = .

a. $\frac{1}{8}$ $\frac{1}{10}$	b. $\frac{4}{9}$ $\frac{1}{2}$	c. $\frac{6}{10}$ $\frac{1}{2}$	d. $\frac{3}{9}$ $\frac{3}{7}$
e. $\frac{4}{7}$ $\frac{6}{13}$	f. $\frac{7}{4}$ $\frac{7}{6}$	g. $\frac{5}{14}$ $\frac{5}{9}$	h. $\frac{4}{20}$ $\frac{2}{20}$
i. $\frac{2}{11}$ $\frac{2}{5}$	j. $\frac{13}{27}$ $\frac{5}{8}$	k. $\frac{12}{24}$ $\frac{1}{2}$	l. $\frac{1}{20}$ $\frac{1}{8}$

A fraction that is more than one (like 6/5) must be bigger than a fraction that is less than one.

$$\frac{6}{5} > \frac{9}{10}$$

In some cases, you might be able to imagine pie pictures in your mind, and "see" which fraction is bigger.

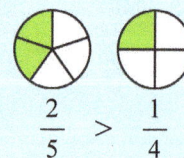

$$\frac{2}{5} > \frac{1}{4}$$

4. Compare the fractions, and write > , < , or = .

a. $\frac{3}{4}$ $\frac{8}{5}$	b. $\frac{8}{7}$ $\frac{3}{3}$	c. $\frac{49}{100}$ $\frac{61}{100}$	d. $\frac{7}{9}$ $\frac{8}{7}$
e. $\frac{9}{10}$ $\frac{3}{4}$	f. $\frac{6}{5}$ $\frac{9}{12}$	g. $\frac{4}{4}$ $\frac{9}{11}$	h. $\frac{1}{3}$ $\frac{3}{9}$

Sometimes none of the "tricks" explained in the previous page work, but we do have one more up our sleeve!

Convert both fractions into like fractions. Then compare.

In the picture on the right, it is hard to be sure if 3/5 is really more than 5/9. Convert both into 45th parts, and then it is easy to see that 27/45 is more than 25/45. Not by much, though!

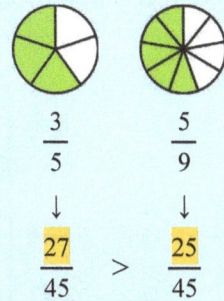

$$\frac{3}{5} \qquad \frac{5}{9}$$
$$\downarrow \qquad \downarrow$$
$$\frac{27}{45} \;>\; \frac{25}{45}$$

5. Convert the fractions into like fractions, and then compare them.

a.	$\dfrac{2}{3}$ $\downarrow$	$\dfrac{5}{8}$ $\downarrow$	b.	$\dfrac{5}{6}$ $\downarrow$	$\dfrac{7}{8}$ $\downarrow$	c.	$\dfrac{1}{3}$ $\downarrow$	$\dfrac{3}{10}$ $\downarrow$	d.	$\dfrac{8}{12}$ $\downarrow$	$\dfrac{7}{10}$ $\downarrow$

| e. | $\dfrac{5}{8}$ $\downarrow$ | $\dfrac{7}{12}$ $\downarrow$ | f. | $\dfrac{11}{8}$ $\downarrow$ | $\dfrac{14}{10}$ $\downarrow$ | g. | $\dfrac{6}{10}$ $\downarrow$ | $\dfrac{58}{100}$ $\downarrow$ | h. | $\dfrac{6}{5}$ $\downarrow$ | $\dfrac{11}{9}$ $\downarrow$ |

| i. | $\dfrac{7}{10}$ $\downarrow$ | $\dfrac{5}{7}$ $\downarrow$ | j. | $\dfrac{43}{100}$ $\downarrow$ | $\dfrac{3}{10}$ $\downarrow$ | k. | $\dfrac{9}{8}$ $\downarrow$ | $\dfrac{8}{7}$ $\downarrow$ | l. | $\dfrac{7}{10}$ $\downarrow$ | $\dfrac{2}{3}$ $\downarrow$ |

6. One cookie recipe calls for ½ cup of sugar. Another one calls for ⅔ cup of sugar.
   Which uses more sugar, a triple batch of the first recipe, or a double batch of the second?

   How much more?

7. Compare the fractions using any method.

a. $\dfrac{5}{12}$	$\dfrac{3}{8}$	**b.** $\dfrac{5}{12}$	$\dfrac{4}{11}$	**c.** $\dfrac{3}{10}$	$\dfrac{1}{5}$	**d.** $\dfrac{3}{8}$	$\dfrac{4}{7}$
e. $\dfrac{4}{15}$	$\dfrac{1}{3}$	**f.** $\dfrac{5}{6}$	$\dfrac{11}{16}$	**g.** $\dfrac{7}{6}$	$\dfrac{10}{8}$	**h.** $\dfrac{5}{12}$	$\dfrac{5}{8}$
i. $\dfrac{3}{4}$	$\dfrac{4}{11}$	**j.** $\dfrac{13}{10}$	$\dfrac{9}{8}$	**k.** $\dfrac{2}{13}$	$\dfrac{1}{5}$	**l.** $\dfrac{1}{10}$	$\dfrac{1}{11}$

8. A coat costs $40. Which is a bigger discount:
   1/4 off the normal price, or 3/10 off the normal price?

   Does your answer change if the original price
   of the coat was $60 instead? Why or why not?

9. Here are three number lines that are divided respectively into halves, thirds, and fifths. Use them to
   help you put the given fractions in order, from the least to the greatest.

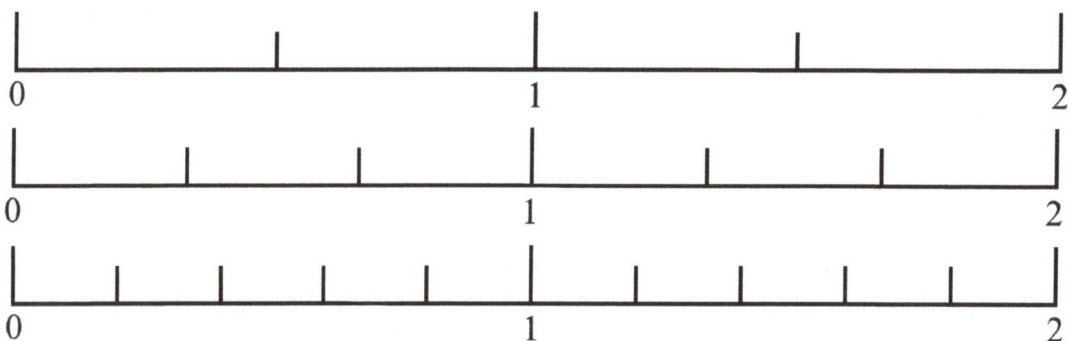

a. $\dfrac{1}{3}, \dfrac{2}{5}, \dfrac{2}{3}, \dfrac{1}{5}, \dfrac{1}{2}$

b. $\dfrac{7}{5}, \dfrac{3}{2}, \dfrac{4}{3}, \dfrac{6}{5}, \dfrac{2}{2}$

_____ < _____ < _____ < _____ < _____   _____ < _____ < _____ < _____ < _____

10. Write the three fractions in order.

a. $\dfrac{7}{8}, \dfrac{9}{10}, \dfrac{7}{9}$	b. $\dfrac{1}{3}, \dfrac{4}{10}, \dfrac{2}{9}$

11. Rebecca made a survey of a group of 600 women. She found that 1/3 of them never exercised, that 22/100 of them swam regularly, 1/5 of them jogged regularly, and the rest were involved in other sports.

   a. Which was a bigger group, the women who jogged or the women who swam?

   b. What fraction of this group of women exercise?

   c. *How many women* in this group exercise?

   d. How many women in this group swim?

The seven dwarfs could not divide a pizza into seven equal slices. The oldest suggested, "Let's cut it into eight slices, let each dwarf have one piece, and give the last piece to the dog."

Then another dwarf said, "No! Let's cut it into 12 slices instead, and give each of us 1 ½ of those pieces, and the dog gets the 1 ½ pieces left over."

Which suggestion would give more pizza to the dog?

Word Problems

Example 1. A recipe calls for 1 ⅓ cups of flour and ½ cup of coconut flakes.
How much in total (in cups) are these two dry ingredients?

Harry added: $1\frac{1}{3} + \frac{1}{2} = 1\frac{2}{5}$, and gave the answer as 1 2/5 cups of dry ingredients.

That is wrong, and you can easily see that, because <u>2/5 is less than ½</u>!

You cannot add ⅓ and ½ and get an answer that is less than ½!

Always check if your answer is reasonable.

1. Cindy needs to make two cakes, one batch of pancakes, and some sauce.
   She needs 3 ½ dl (*deciliters*) of flour for one cake, 5 dl of flour for a batch of pancakes,
   and ¾ dl of flour for the sauce.

   a. How much flour does she need in total?

   b. A 1-kilogram bag of flour is about 15 deciliters.
   Will one bag of flour be enough for her to make the three recipes?

2. Lily's notebook is 3 ¼ inches wide and 6 ⅛ inches long.
   She wants to glue a picture on the front so that
   the margins on all sides are ¾ inches.

   What size should the picture be (width and length)?

picture

3. A company divided a project so that Mark would do 1/10 of it, Leslie would do 1/2 of it, and Jerry the rest. What part (fraction) of the job was left for Jerry to do?

4. Bob earned $4,000. He paid 19/100 of this in taxes and 2/10 of it to pay back a loan.

   a. What part (what fraction) of his money does he have left after those payments?

   b. Calculate how many dollars Bob has left.
   (Hint: A bar model can help.)

5. Mom made a full pitcher of smoothie for the family. She gave ⅛ of it to Anna, ⅛ of it to Jack, ¼ of it to Dad, and drank ¼ of it herself.

   a. What fraction of the total smoothie is left in the pitcher now?

   b. If the original amount of smoothie was 2,000 ml, how much of it is left now?
   (Hint: A bar model can help.)

Measuring in Inches

Here are four rulers that all measure in inches. They are *not* to scale. Instead, they are magnified to be bigger than in real life, so you can see how they are divided into parts.

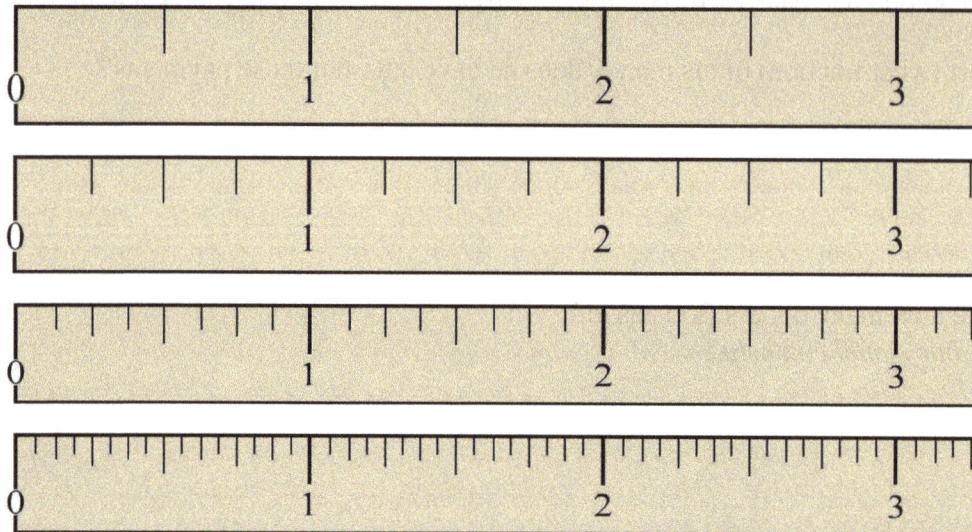

1. Locate the ½-inch mark, 1 ½ -inch mark, and 2 ½-inch mark on all of the rulers above.

2. Locate the ¼-inch and ¾-inch marks for the bottom three rulers: between 0 and 1 inches, between 1 and 2 inches, and between 2 and 3 inches.

3. The ruler below has tick marks for every eighth of an inch. Find and label the marks for these measurements: 1/8 inch, 5/8 inch, 7/8 inch, 1 5/8 inches, and 2 3/8 inches.

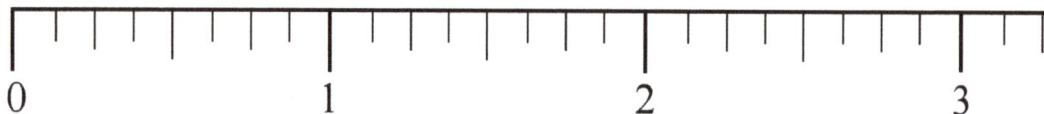

Now locate these *same* points on the ruler that measures in 16th parts of an inch:

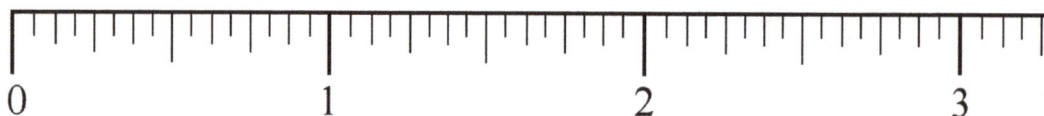

4. On the ruler that measures in 16th parts of an inch, locate tick marks for these points:

- 3/16 inch
- 7/16 inch
- 11/16 inch

- 1 1/8 inches
- 2 3/8 inches
- 7/8 inch

- 1/4 inch
- 1 1/4 inches
- 2 3/4 inches

5. Measure the colored lines with the given rulers. If the end of the line does not fall exactly on a tick mark, then read the mark that is CLOSEST to the end of the line.

a.

b.

c.

d.

e.

f.

g.

h.

i.

6. Measure the length of six items in your home or classroom, using a 1/16-inch ruler, and write the results below. (You can cut out a ruler from the following page.)

a. _____ _____ in.   b. _____ _____ in.

c. _____ _____ in.   d. _____ _____ in.

e. _____ _____ in.   f. _____ _____ in.

7. Measure the following lines using different rulers. Most student rulers are 1/16-inch rulers. You can cut out rulers from the bottom of this page, and tape them on top of an existing ruler, to get a 1/4-inch and 1/8-inch ruler.

What do you notice about the three measurements for each line?

a. Using the 1/4-inch ruler: _____ in. Using the 1/8-inch ruler: _____ in. Using the 1/16-inch ruler: _____ in.	**b.** Using the 1/4-inch ruler: _____ in. Using the 1/8-inch ruler: _____ in. Using the 1/16-inch ruler: _____ in.
c. Using the 1/4-inch ruler: _____ in. Using the 1/8-inch ruler: _____ in. Using the 1/16-inch ruler: _____ in.	**d.** Using the 1/4-inch ruler: _____ in. Using the 1/8-inch ruler: _____ in. Using the 1/16-inch ruler: _____ in.
e. Using the 1/4-inch ruler: _____ in. Using the 1/8-inch ruler: _____ in. Using the 1/16-inch ruler: _____ in.	**f.** Using the 1/4-inch ruler: _____ in. Using the 1/8-inch ruler: _____ in. Using the 1/16-inch ruler: _____ in.

These cut-out rulers can be placed on top of an existing ruler.

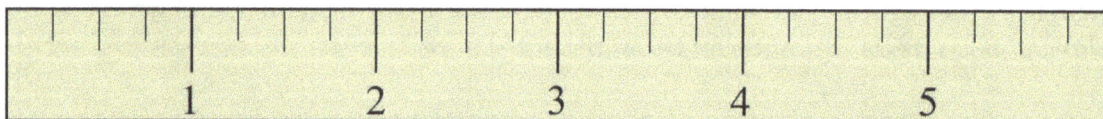

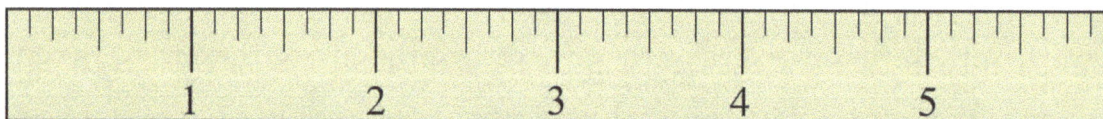

Line Plots and More Measuring

This is a **line plot**. It is a type of graph that uses a number line to display data. This line plot shows how much time Kate spent for homework in nine different days.

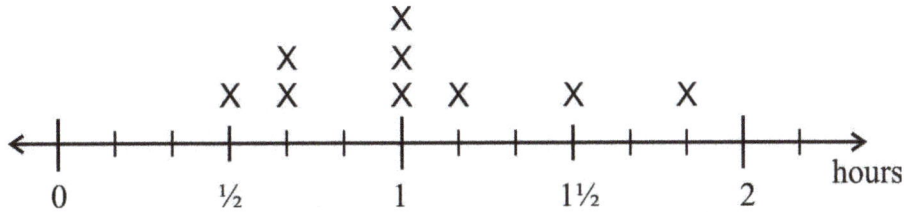

Each **x**-mark signifies a particular item or value from the data. When several data items have the same value, their **x**-marks are stacked.

1. Answer the questions based on the line plot above.

   a. On how many days did Kate spend exactly 1 hour on homework?

   b. What is the longest amount of time she spent on homework within these nine days?

   c. How much time in total did she spend on homework within these nine days?

2. Janet checked the amount of sugar in 10 different cookie recipes. The amounts, in cups, were:

   1 ½   1 ⅜   1   1 ¾   1 ½   1 ⅛   1 ¼   1 ¼   1 ½   ¾

   a. Make a **line plot** by drawing an X-mark for each measurement above the number line.

   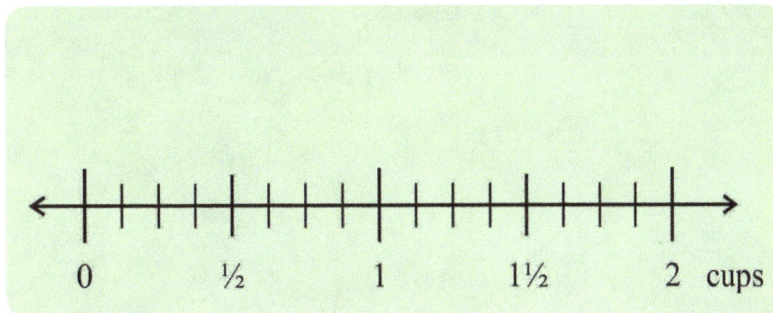

   b. If Janet made the recipe with the least amount of sugar three times, how much sugar would she need?

   c. If Janet made the recipe with the largest amount of sugar three times, how much sugar would she need?

3. These are the lengths of cockroaches in Jake's collection (in inches):

1 ¼   1 ⅛   1 ⅛   1 ½   1   1 ⅛   1 ⅜   1 ¾   1 ⅜   ⅞   1 ¼   2 ⅛   ½

1 ¼   1 ¼   1 ½   1 ½   1 ½   1 ⅝

a. Make a line plot. This time, you will need to do the scaling on the number line.

⟵───────────────────────────────────⟶

b. Are there more cockroaches that are strictly less than 1 ½ inch long, or those that are 1 ½ inch long or more?

c. If Jake puts the three longest cockroaches end-to-end, how long a "train" do they make?

4. **a.** Carefully measure the sides of the quadrilateral below, in inches. Mark the length of each side next to the side.

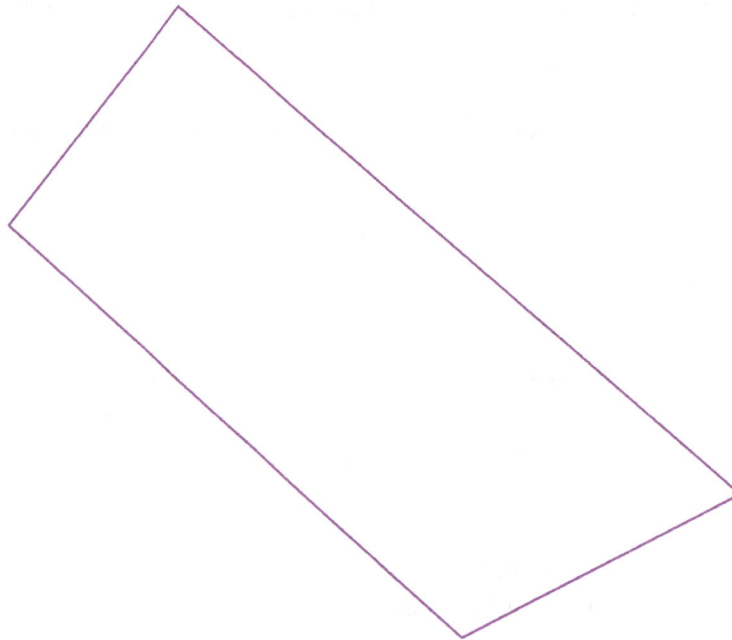

b. Find the perimeter of the quadrilateral.

5. The dimensions (width and length) of one smartphone are 3 ⅛ in by 6 ¼ in, and of another are 3 in by 6 ⅛ in. How much larger perimeter does the first phone have than the second?

6. Measure a bunch of pencils to the nearest 1/16 of an inch. Then make a line plot of your data.

My data:

Review

1. Write as fractions. Think of the shortcut.

a. $9\frac{1}{2}$	**b.** $5\frac{6}{11}$	**c.** $8\frac{2}{7}$	**d.** $5\frac{6}{100}$

2. Write as mixed numbers.

a. $\frac{41}{10}$	**b.** $\frac{19}{3}$	**c.** $\frac{28}{9}$	**d.** $\frac{32}{12}$

3. Rewrite the division problem $23 \div 6 = 3\,R5$ as a problem where a fraction is changed to a mixed number.

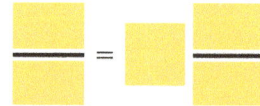

$$\frac{\square}{\square} = \square\frac{\square}{\square}$$

4. Subtract. Regroup if necessary. Check that your answer is reasonable.

a. $9\frac{4}{8}$ $-3\frac{7}{8}$	**b.** $12\frac{3}{20}$ $-5\frac{11}{20}$	**c.** $10\frac{3}{5}$ $-5\frac{1}{3}$

5. Add and subtract. Check that your answer is reasonable.

a. $\frac{5}{7} + \frac{1}{3}$	**b.** $\frac{3}{10} + \frac{1}{3}$
c. $3\frac{2}{7} - 1\frac{6}{7}$	**d.** $2\frac{4}{5} + 3\frac{1}{4}$

6. Compare the fractions, and write <, >, or = in the box.

a. $\dfrac{1}{2}$ ☐ $\dfrac{3}{5}$	b. $\dfrac{3}{11}$ ☐ $\dfrac{1}{3}$	c. $\dfrac{7}{10}$ ☐ $\dfrac{70}{100}$	d. $\dfrac{1}{4}$ ☐ $\dfrac{28}{100}$
e. $\dfrac{2}{3}$ ☐ $\dfrac{8}{9}$	f. $\dfrac{1}{4}$ ☐ $\dfrac{2}{15}$	g. $\dfrac{21}{16}$ ☐ $\dfrac{25}{16}$	h. $\dfrac{5}{11}$ ☐ $\dfrac{1}{2}$

7. Betty uses 3 ⅛ feet of material to make one shirt. She has one piece
   that is 5 ½ feet and another piece that is 4 ¼ feet. She made one
   shirt from *each* piece of material.
   How much material does she have left now, in total?

8. Of a piece of land, 32/100 is planted in wheat, 42/100 is planted
   in barley, 2/10 is planted in oats, and the remainder is resting.
   What part (fraction) of the land is resting?

9. Which is a better deal: 1/5 off of a book that costs $35,
   or 2/11 off of a book that costs $33?

   Would the situation change if both deals involved a book
   that costs $50? Explain.

10. Twelve beakers have various amounts of oil in them. The line plot shows how much oil each beaker has, in cups.

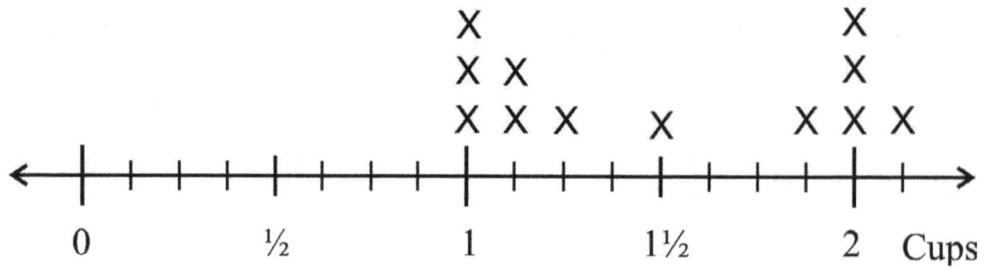

```
                        X                       X
                        X  X                    X
                        X  X  X        X        X  X  X
     ←——┼——┼——┼——┼——┼——┼——┼——┼——┼——┼——┼——┼——┼——┼——┼——→
        0        ½        1        1½        2   Cups
```

a. How many beakers have 1 ⅛ cups of oil in them?

b. How many have 1 ⅞ cups?

c. How much oil in total is in the five
   beakers that have the most oil?

Answer Key

Review: Mixed Numbers, pp. 8-10

1. a. 1 1/3   b. 2 2/6   c. 3 3/5

2.

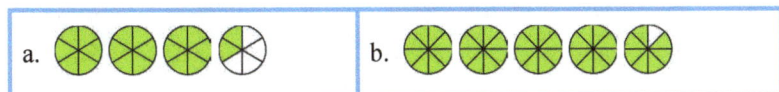

3. a. 2 2/5   b. 1 6/7

4. a. 2/4   b. 1 1/4   c. 3 3/4   d. 4 2/4

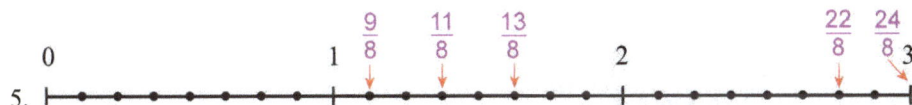

5.

Mixed numbers to fractions

Example 1. To write $4\frac{2}{9}$ as a fraction, we *count* all the ninths:

- Each pie has nine ninths, so the four complete pies have $4 \times \underline{\,9\,} = \underline{\,36\,}$ ninths.

- Additionally, the incomplete pie has $\underline{\,2\,}$ ninths.

- The total is $\underline{\,38\,}$ ninths or $\frac{38}{9}$.

---

6. a. 1 2/5 = 7/5   b. 2 4/6 = 16/6   c. 2 3/8 = 19/8
   d. 4 5/12 = 53/12   e. 3 1/4 = 13/4

7. There are 5 whole pies, and each pie has 13 slices. So 5 × 13 tells us the number of slices in the whole pies.
   Then the fractional part 9/13 means that we add 9 slices to that. In total we get 74 slices, and each one

   is a 13th part. So the fraction is $\frac{74}{13}$.

8. a. 15/2   b. 20/3   c. 75/9   d. 66/10
   e. 27/11   f. 97/12   g. 37/16   h. 39/8

9. How many 7s are there in 58? <u>8</u>
   After that, how many "slices" are left over? <u>2</u>
   What math operation helps you with the above? <u>division</u>
   58/7 = 8 2/7

10.

a. $47 \div 4 = 11$ R3	b. $35 \div 8 = 4$ R3	c. $19 \div 2 = 9$ R1
$\frac{47}{4} = 11\frac{3}{4}$	$\frac{35}{8} = 4\frac{3}{8}$	$\frac{19}{2} = 9\frac{1}{2}$
d. $35 \div 6 = 5$ R5	e. $72 \div 10 = 7$ R2	f. $22 \div 7 = 3$ R1
$\frac{35}{6} = 5\frac{5}{6}$	$\frac{72}{10} = 7\frac{2}{10}$	$\frac{22}{7} = 3\frac{1}{7}$

11. a. 7 6/8   b. 5 1/3   c. 5 2/5   d. 3 5/9
    e. 3 1/2   f. 6 1/4   g. 8 2/6   h. 6 2/5
    i. 2 2/11   j. 13   k. 7 1/8   l. 9 6/9

53

Adding Mixed Numbers, pp. 11-13

1. The student added the numerators separately, and the denominators separately. The answer 7/16, is not correct, and not even reasonable (because it's actually less than 1/2)!

  The correct answer is 7/8.

2. a. 8/10   b. 1 4/8   c. 1   d. 1 3/4   e. 1 4/5

3. a. 10 3/4   b. 18 8/9   c. 13 3/4   d. 32 10/12

4. b. $1\frac{11}{9} = 2\frac{2}{9}$ c. $6\frac{7}{4} = 7\frac{3}{4}$ d. $3\frac{13}{8} = 4\frac{5}{8}$

5.

a. $2\frac{5}{6} + 1\frac{5}{6} = 4\frac{4}{6}$	b. $2\frac{1}{7} + \frac{3}{7} + \frac{4}{7} = 3\frac{1}{7}$

6. a. 12   b. 6 2/5   c. 8 1/9

7. The perimeter is 7 3/8 + 5 7/8 + 3 4/8 = 15 14/8 = 16 6/8 inches.

8.

a.	b.	c.	d.
$4\frac{3}{7}$	$3\frac{3}{5}$	$4\frac{6}{9}$	$7\frac{6}{8}$
$+\ 5\frac{5}{7}$	$+\ 3\frac{4}{5}$	$+\ 2\frac{7}{9}$	$+\ 2\frac{7}{8}$
$9\frac{8}{7} = 10\frac{1}{7}$	$6\frac{7}{5} = 7\frac{2}{5}$	$6\frac{13}{9} = 7\frac{4}{9}$	$9\frac{13}{8} = 10\frac{5}{8}$

9. a. 3 3/4   b. 4 1/3

10. a. 1 2/4   b. 2 1/5

11. 3 13/6 = 5 1/6. You get two additional "pies" out of 13/6, plus 1/6.

12. a. 6   b. 11 3/5
    c. 10 1/8   d. 8 3/10

13. Jeremy runs 4 × 2 ¼ miles = 2 ¼ miles + 2 ¼ miles + 2 ¼ miles + 2 ¼ miles = 9 miles in a week.
    Robert runs 3 × 3 ½ miles = 3 ½ miles + 3 ½ miles + 3 ½ miles = 10 ½ miles in a week.
    <u>Robert runs 1 ½ miles more in a week.</u>

14.

a.	b.	c.	d.
$10\frac{7}{9}$	$1\frac{5}{11}$	$2\frac{5}{6}$	$1\frac{7}{10}$
$2\frac{5}{9}$	$3\frac{9}{11}$	$5\frac{4}{6}$	$\frac{9}{10}$
$+\ 3\frac{8}{9}$	$+\ 2\frac{8}{11}$	$+\ 2\frac{3}{6}$	$+\ 10\frac{6}{10}$
$15\frac{20}{9} = 17\frac{2}{9}$	$6\frac{22}{11} = 8$	$9\frac{12}{6} = 11$	$11\frac{22}{10} = 13\frac{2}{10}$

Subtracting Mixed Numbers 1, pp. 14-17

1. a. 1 7/6   b. 2 9/8   c. 1 11/9   d. 1 8/5   e. 2 13/10   f. 1 5/4

2.

a. $4\frac{2}{9} - 1\frac{8}{9}$ ↓ $= 3\frac{11}{9} - 1\frac{8}{9} = 2\frac{3}{9}$	b. $5\frac{3}{12} - 2\frac{7}{12}$ ↓ $= 4\frac{15}{12} - 2\frac{7}{12} = 2\frac{8}{12}$
c. $5\frac{7}{10} - 3\frac{9}{10}$ ↓ $= 4\frac{17}{10} - 3\frac{9}{10} = 1\frac{8}{10}$	d. $4\frac{3}{8} - 1\frac{7}{8}$ ↓ $= 3\frac{11}{8} - 1\frac{7}{8} = 2\frac{4}{8}$

3.

a. $\cancel{3}\ \overset{2}{\cancel{}}\ \frac{\cancel{4}}{9}\rightarrow 2\frac{13}{9}$ $-\ \ \frac{8}{9}$ $2\ \frac{5}{9}$	b. $\cancel{7}\ \frac{\cancel{4}}{9}\rightarrow 6\frac{13}{9}$ $-\ 2\ \frac{7}{9}$ $4\ \frac{6}{9}$	c. $\cancel{12}\ \frac{\cancel{9}}{12}\rightarrow 11\frac{21}{12}$ $-\ 6\ \frac{11}{12}$ $5\ \frac{10}{12}$	d. $\cancel{8}\ \frac{\cancel{3}}{14}\rightarrow 7\frac{17}{14}$ $-\ 5\ \frac{9}{14}$ $2\ \frac{8}{14}$
e. $14\ \frac{7}{9}$ $-\ 3\ \frac{5}{9}$ $11\ \frac{2}{9}$	f. $\cancel{11}\ \frac{\cancel{5}}{21}\rightarrow 10\frac{26}{21}$ $-\ 7\ \frac{15}{21}$ $3\ \frac{11}{21}$	g. $\cancel{26}\ \frac{\cancel{4}}{19}\rightarrow 25\frac{23}{19}$ $-\ 14\ \frac{15}{19}$ $11\ \frac{8}{19}$	h. $\cancel{10}\ \frac{\cancel{3}}{20}\rightarrow 9\frac{23}{20}$ $-\ 5\ \frac{7}{20}$ $4\ \frac{16}{20}$

4. a. 1 3/6   b. 3/5   c. 3/7

5. Ellie has 4 yd − 7/8 yd − 7/8 yd = <u>2 2/8 yd</u> material left.

6.

a. $3\frac{1}{9} - 1\frac{5}{9}$ $= 3\frac{1}{9} - 1\frac{1}{9} - \frac{4}{9}$ $= \quad 2 \quad - \quad \frac{4}{9} = 1\frac{5}{9}$	b. $2\frac{5}{12} - 1\frac{11}{12}$ $= 2\frac{5}{12} - 1\frac{5}{12} - \frac{6}{12}$ $= \quad 1 \quad - \quad \frac{6}{12} = \frac{6}{12}$

7. a. reasonable   b. not reasonable; 2 3/8   c. not reasonable; 2 9/13
   d. not reasonable; 4 11/15   e. reasonable   f. not reasonable; 3 85/100

8. The third side of the triangle is 10 1/8 in − 3 5/8 in − 3 5/8 in = <u>2 7/8 in</u>.

9. The two recipes call for 1 3/4 C + 1 3/4 C = 3 2/4 C − 3/4 C = 2 3/4 C. So, he needs <u>2 3/4 cups more</u>.

Puzzle Corner: Yes. Nine times 1 5/9 is 14, so you can subtract 1 5/9 nine times, starting from 14, and reach zero.

Subtracting Mixed Numbers 2, pp. 18-19

1. a. 3 4/8   b. 3 9/15
   c. 3 2/30   d. 11 6/12

2. You have 3 ¾ − ¾ − ¾ = 2 ¼ kg of beef left.

3. a. 1 10/11   b. 5 1/7   c. 4 8/15

4. a. 5 3/5   b. 5 11/12   c. 3 7/9

5. a. 4 2/4   b. 9 4/6
   c. 8 4/8   d. 10 4/12

6.

m. $4\frac{11}{12}$	i. $1\frac{2}{10}$	e. $12\frac{6}{9}$	a. $2\frac{4}{9}$
k. $3\frac{7}{11}$			c. $2\frac{4}{11}$
b. $2\frac{12}{15}$		h. 2	l. $6\frac{6}{8}$
d. $1\frac{7}{9}$	f. $1\frac{3}{11}$	g. $5\frac{6}{12}$	j. $5\frac{4}{8}$

Equivalent Fractions 1, pp. 20-22

1.

2.

$$\frac{1}{2} = \frac{2}{4} = \frac{3}{6} = \frac{4}{8} = \frac{5}{10} = \frac{6}{12} = \frac{7}{14} = \frac{8}{16}$$

Equivalent Fractions 1, cont.

3.

a. Split each piece <u>in two</u>.	b. Split each piece <u>into three</u>.	c. Split each piece <u>in two</u>.
×2 $\frac{2}{5} = \frac{4}{10}$ ×2	×3 $\frac{1}{2} = \frac{3}{6}$ ×3	×2 $\frac{2}{3} = \frac{4}{6}$ ×2
d. Split each piece <u>in two</u>.	e. Split each piece <u>into three</u>.	f. Split each piece <u>in two</u>.
×2 $\frac{1}{4} = \frac{2}{8}$ ×2	×3 $\frac{3}{3} = \frac{9}{9}$ ×3	×2 $\frac{1}{5} = \frac{2}{10}$ ×2
g. Split each piece <u>in two</u>.	h. Split each piece <u>in two</u>.	i. Split each piece <u>into five</u>.
×2 $\frac{1}{2} = \frac{2}{4}$ ×2	×2 $\frac{3}{8} = \frac{6}{16}$ ×2	×5 $\frac{1}{2} = \frac{5}{10}$ ×5

4.

a.	b.	c.	d.	e.
$\frac{3}{4} = \frac{12}{16}$	$\frac{5}{8} = \frac{10}{16}$	$\frac{1}{2} = \frac{6}{12}$	$\frac{2}{7} = \frac{8}{28}$	$\frac{1}{4} = \frac{5}{20}$
f.	g.	h.	i.	j.
$\frac{2}{7} = \frac{6}{21}$	$\frac{5}{8} = \frac{50}{80}$	$\frac{1}{2} = \frac{8}{16}$	$\frac{3}{5} = \frac{21}{35}$	$\frac{3}{7} = \frac{24}{56}$

5.

a. Pieces were split into <u>three</u>.	b. Pieces were split into <u>four</u>.	c. Pieces were split into <u>three</u>.	d. Pieces were split into <u>two</u>.	e. Pieces were split into <u>four</u>.
×3 $\frac{4}{7} = \frac{12}{21}$ ×3	×4 $\frac{4}{5} = \frac{16}{20}$ ×4	×3 $\frac{1}{6} = \frac{3}{18}$ ×3	×2 $\frac{6}{7} = \frac{12}{14}$ ×2	×4 $\frac{2}{3} = \frac{8}{12}$ ×4
f. $\frac{3}{10} = \frac{9}{30}$	g. $\frac{2}{11} = \frac{6}{33}$	h. $\frac{4}{7} = \frac{32}{56}$	i. $\frac{1}{6} = \frac{9}{54}$	j. $\frac{7}{8} = \frac{56}{64}$

Equivalent Fractions 1, cont.

6. a. 2/3 = 8/12 = 16/24

   b. 5/6 = 10/12 = 20/24

    c. Answers will vary; any of the following will do:
1/12, 2/12, 3/12, 5/12, 6/12, 7/12, 9/12, 10/12, or 11/12.

    d. Answers will vary; any of the following will do:
1/24, 3/24, 5/24, 7/24, 9/24, 11/24, 13/24, 15/24,
17/24, 19/24, 21/24, or 23/24.

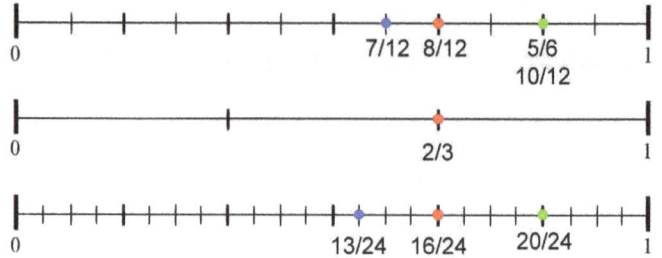

7. a. & b.

	Fraction	Equivalent fraction
Dad	1/2	6/12
Mom	1/3	4/12
Cindy	1/3	4/12
Derek		10/12

   b. Derek ate 10/12 of a pizza. Note: the total comes to 24/12, which is equal to two pizzas.

Equivalent Fractions 2, pp. 23-24

1.

a. Split each slice into three.

$$\times 3$$
$$\frac{7}{4} = \frac{21}{12}$$
$$\times 3$$

b. Split each slice in two.

$$\times 2$$
$$\frac{12}{5} = \frac{24}{10}$$
$$\times 2$$

c. Split each slice in two.

$$\times 2$$
$$\frac{5}{3} = \frac{10}{6}$$
$$\times 2$$

d. Split each slice into four.

$$\times 4$$
$$\frac{4}{2} = \frac{16}{8}$$
$$\times 4$$

2. a. 5 56/80   b. 5 28/40   c. 63/28   d. 42/6
   e. 6 12/54   f. 2 21/28   g. 32/10   h. 40/15

3.

whole pies	halves	thirds	fourths	fifths	tenths	hundredths
$\frac{3}{1}$	$\frac{6}{2}$	$\frac{9}{3}$	$\frac{12}{4}$	$\frac{15}{5}$	$\frac{30}{10}$	$\frac{300}{100}$

4.

halves	fourths	sixths	eighths	tenths	twentieths	hundredths
$\frac{5}{2}$	$\frac{10}{4}$	$\frac{15}{6}$	$\frac{20}{8}$	$\frac{25}{10}$	$\frac{50}{20}$	$\frac{250}{100}$

Equivalent Fractions 2, cont.

5.

a. $\frac{5}{7} = \frac{20}{28}$ The pieces were split into _4_ .	b. NOT POSSIBLE	c. NOT POSSIBLE	d. $\frac{2}{3} = \frac{8}{12}$ The pieces were split into _4_ .	e. NOT POSSIBLE
f. NOT POSSIBLE	g. $\frac{2}{9} = \frac{14}{63}$ The pieces were split into _7_ .	h. $\frac{5}{4} = \frac{40}{32}$ The pieces were split into _8_ .	i. $\frac{1}{3} = \frac{5}{15}$ The pieces were split into _5_ .	j. NOT POSSIBLE

6. Answers will vary. For example:

If the new numerator is not divisible by the old one, or if the new denominator is not divisible by the old one, then the conversion is not possible.
Or, if the numerator does not "go into" or divide into the new numerator and similarly with the denominators, then we cannot find an equivalent fraction.

7.

a. $\frac{3}{4} = \frac{6}{8} = \frac{9}{12} = \frac{12}{16} = \frac{15}{20} = \frac{18}{24} = \frac{21}{28} = \frac{24}{32} = \frac{27}{36}$

b. $\frac{5}{3} = \frac{10}{6} = \frac{15}{9} = \frac{20}{12} = \frac{25}{15} = \frac{30}{18} = \frac{35}{21} = \frac{40}{24} = \frac{45}{27}$

Adding and Subtracting Unlike Fractions, pp. 25-27

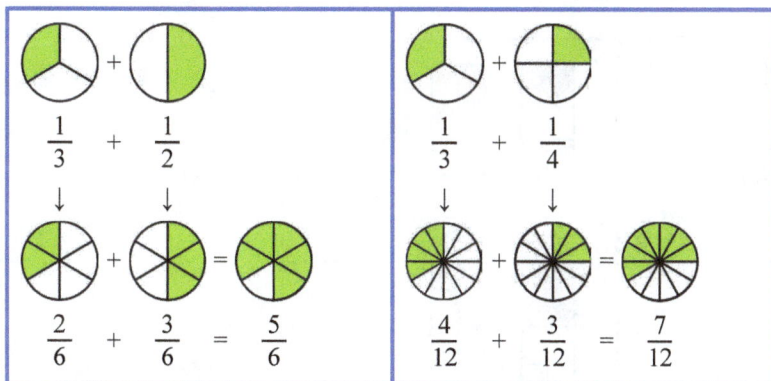

$\frac{1}{3} + \frac{1}{2}$
↓   ↓
$\frac{2}{6} + \frac{3}{6} = \frac{5}{6}$

$\frac{1}{3} + \frac{1}{4}$
↓   ↓
$\frac{4}{12} + \frac{3}{12} = \frac{7}{12}$

1.

a. $\frac{1}{2} + \frac{1}{4}$ ↓ ↓ $\frac{2}{4} + \frac{1}{4} = \frac{3}{4}$	b. $\frac{2}{5} + \frac{1}{2}$ ↓ ↓ $\frac{4}{10} + \frac{5}{10} = \frac{9}{10}$	c. $\frac{3}{9} + \frac{1}{3}$ ↓ ↓ $\frac{3}{9} + \frac{3}{9} = \frac{6}{9}$

2.

a. $\dfrac{1}{2} + \dfrac{1}{6}$ $\downarrow \quad \downarrow$ $\dfrac{3}{6} + \dfrac{1}{6} = \dfrac{4}{6}$	b. $\dfrac{1}{8} + \dfrac{1}{4}$ $\downarrow \quad \downarrow$ $\dfrac{1}{8} + \dfrac{2}{8} = \dfrac{3}{8}$	c. $\dfrac{1}{6} + \dfrac{1}{4}$ $\downarrow \quad \downarrow$ $\dfrac{2}{12} + \dfrac{3}{12} = \dfrac{5}{12}$
d. $\dfrac{5}{6} - \dfrac{1}{2}$ $\downarrow \quad \downarrow$ $\dfrac{5}{6} - \dfrac{3}{6} = \dfrac{2}{6}$	e. $\dfrac{5}{8} - \dfrac{1}{4}$ $\downarrow \quad \downarrow$ $\dfrac{5}{8} - \dfrac{2}{8} = \dfrac{3}{8}$	f. $\dfrac{5}{6} - \dfrac{1}{4}$ $\downarrow \quad \downarrow$ $\dfrac{10}{12} - \dfrac{3}{12} = \dfrac{7}{12}$

3.

a. $\dfrac{1}{2} + \dfrac{1}{8}$ $\downarrow \quad \downarrow$ $\dfrac{4}{8} + \dfrac{1}{8} = \dfrac{5}{8}$	b. $\dfrac{3}{10} + \dfrac{1}{5}$ $\downarrow \quad \downarrow$ $\dfrac{3}{10} + \dfrac{2}{10} = \dfrac{5}{10}$	c. $\dfrac{2}{5} + \dfrac{1}{2}$ $\downarrow \quad \downarrow$ $\dfrac{4}{10} + \dfrac{5}{10} = \dfrac{9}{10}$
d. $\dfrac{1}{2} + \dfrac{3}{8}$ $\downarrow \quad \downarrow$ $\dfrac{4}{8} + \dfrac{3}{8} = \dfrac{7}{8}$	e. $\dfrac{9}{10} - \dfrac{2}{5}$ $\downarrow \quad \downarrow$ $\dfrac{9}{10} - \dfrac{4}{10} = \dfrac{5}{10}$	f. $\dfrac{4}{5} - \dfrac{1}{2}$ $\downarrow \quad \downarrow$ $\dfrac{8}{10} - \dfrac{5}{10} = \dfrac{3}{10}$

4.

a. $\dfrac{2}{4} + \dfrac{3}{4} = \dfrac{5}{4} = 1\dfrac{1}{4}$	b. $\dfrac{4}{8} + \dfrac{5}{8} = \dfrac{9}{8} = 1\dfrac{1}{8}$	c. $\dfrac{2}{6} + \dfrac{5}{6} = \dfrac{7}{6} = 1\dfrac{1}{6}$
d. $\dfrac{5}{10} + \dfrac{4}{10} = \dfrac{9}{10}$	e. $\dfrac{5}{15} + \dfrac{6}{15} = \dfrac{11}{15}$	f. $\dfrac{2}{6} + \dfrac{3}{6} = \dfrac{5}{6}$

5.

Types of parts:	Converted to:	Types of parts:	Converted to:
2nd parts and 4th parts	4th parts	2nd parts and 5th parts	10th parts
2nd parts and 8th parts	8th parts	3rd parts and 5th parts	15th parts
3rd parts and 6th parts	6th parts	3rd parts and 2nd parts	6th parts

b. The two denominators always "go into" the number that tells us what kind of parts we are converting to. In other words, we need to find a number that is divisible by the two denominators, or in yet other words, a number that is a multiple of both of the denominators.

Adding and Subtracting Unlike Fractions, cont.

6.

a. $\frac{1}{2} + \frac{2}{3}$	b. $\frac{2}{3} - \frac{2}{5}$	c. $\frac{1}{3} + \frac{3}{4}$
$\downarrow$ $\downarrow$	$\downarrow$ $\downarrow$	$\downarrow$ $\downarrow$
$\frac{3}{6} + \frac{4}{6} = \frac{7}{6} = 1\frac{1}{6}$	$\frac{10}{15} - \frac{6}{15} = \frac{4}{15}$	$\frac{4}{12} + \frac{9}{12} = \frac{13}{12} = 1\frac{1}{12}$

Finding the (Least) Common Denominator, pp. 28-30

1.

a. $\frac{1}{3} + \frac{3}{5}$	b. $\frac{6}{7} - \frac{1}{2}$	c. $\frac{1}{6} + \frac{2}{5}$
$\downarrow$ $\downarrow$	$\downarrow$ $\downarrow$	$\downarrow$ $\downarrow$
$\frac{5}{15} + \frac{9}{15} = \frac{14}{15}$	$\frac{12}{14} - \frac{7}{14} = \frac{5}{14}$	$\frac{5}{30} + \frac{12}{30} = \frac{17}{30}$
d. $\frac{5}{9} - \frac{1}{3}$	e. $\frac{1}{8} + \frac{3}{4}$	f. $\frac{5}{7} - \frac{2}{3}$
$\downarrow$ $\downarrow$	$\downarrow$ $\downarrow$	$\downarrow$ $\downarrow$
$\frac{5}{9} - \frac{3}{9} = \frac{2}{9}$	$\frac{1}{8} + \frac{6}{8} = \frac{7}{8}$	$\frac{15}{21} - \frac{14}{21} = \frac{1}{21}$
g. $\frac{2}{5} + \frac{1}{4}$	h. $\frac{5}{6} - \frac{3}{4}$	i. $\frac{3}{4} - \frac{3}{7}$
$\downarrow$ $\downarrow$	$\downarrow$ $\downarrow$	$\downarrow$ $\downarrow$
$\frac{8}{20} + \frac{5}{20} = \frac{13}{20}$	$\frac{10}{12} - \frac{9}{12} = \frac{1}{12}$	$\frac{21}{28} - \frac{12}{28} = \frac{9}{28}$

2. a. 20

    b. 21

    c. 10 (20 is okay, too, though not the best)

    d. 12 (24 and 48 are okay too, though not the best)

    e. 14

    f. 18 (36 and 54 are okay too, though not the best)

3.

a. $\frac{4}{5} + \frac{1}{4}$	b. $\frac{2}{3} - \frac{1}{7}$	c. $\frac{3}{10} + \frac{1}{2}$
$\downarrow$ $\downarrow$	$\downarrow$ $\downarrow$	$\downarrow$ $\downarrow$
$\frac{16}{20} + \frac{5}{20} = \frac{21}{20}$	$\frac{14}{21} - \frac{3}{21} = \frac{11}{21}$	$\frac{3}{10} + \frac{5}{10} = \frac{8}{10}$
d. $\frac{4}{12} + \frac{1}{4}$	e. $\frac{1}{2} - \frac{2}{7}$	f. $\frac{5}{6} - \frac{4}{9}$
$\downarrow$ $\downarrow$	$\downarrow$ $\downarrow$	$\downarrow$ $\downarrow$
$\frac{4}{12} + \frac{3}{12} = \frac{7}{12}$	$\frac{7}{14} - \frac{4}{14} = \frac{3}{14}$	$\frac{15}{18} - \frac{8}{18} = \frac{7}{18}$

Finding the (Least) Common Denominator, cont.

4. a. The LCD is 24. 10/24 + 9/24 = 19/24
   b. The LCD is 44. 77/44 − 36/44 = 41/44
   c. The LCD is 36. 3/36 + 4/36 = 7/36
   d. The LCD is 72. 63/72 − 32/72 = 31/72

Add and Subtract: More Practice, pp. 31-33

1.

$\frac{1}{10} + \frac{8}{9} = \cancel{\frac{20}{45}}$ $\frac{7}{12} + \frac{1}{18} = \frac{23}{36}$ $\frac{7}{10} + \frac{8}{9} = \frac{143}{90}$ $\frac{5}{8} + \frac{9}{16} = \cancel{\frac{14}{32}}$

$\frac{2}{11} + \frac{1}{10} = \cancel{\frac{91}{110}}$ $\frac{5}{12} - \frac{2}{5} = \cancel{\frac{43}{60}}$ $\frac{4}{5} - \frac{1}{13} = \frac{47}{65}$ $\frac{8}{15} + \frac{5}{6} = \cancel{\frac{12}{3}}$

2.

| a. $\frac{2}{13} + \frac{15}{16}$ | ☐ $\frac{17}{208}$ ☐ $\frac{129}{208}$ 🟧 $\frac{227}{208}$ | b. $\frac{7}{8} - \frac{1}{9}$ | 🟧 $\frac{55}{72}$ ☐ $\frac{79}{72}$ ☐ $\frac{12}{72}$ | c. $\frac{5}{12} - \frac{1}{8}$ | ☐ $\frac{79}{96}$ 🟧 $\frac{28}{96}$ ☐ $\frac{56}{96}$ |

3.

a. Amanda added: $\frac{5}{8} + \frac{5}{8} = \frac{10}{16}$

How can you tell it is wrong?

Answers will vary. For example:

The answer, 10/16, is equal to 5/8!

Or, we are adding like fractions, which is easy: you simply add the number of pieces. We have 5 and 5 pieces, or 10 pieces, so the answer is 10 eighths.

Correct answer: 10/8 = 1 2/8 = 1 1/4.

b. Robert subtracted: $\frac{7}{9} - \frac{1}{2} = \frac{6}{7}$

How can you tell it is wrong?

Answers will vary. For example:

7/9 is nearly 1, and we subtract one half from it. The answer, 6/7, is also nearly 1, so it does not make sense.

Correct answer: 14/18 − 9/18 = 5/18.

4.

Olivia subtracted: $\frac{7}{12} - \frac{1}{4} = \frac{1}{2}$ Correct answer: 7/12 − 3/12 = 4/12 (which equals 1/3).

How can you tell it is wrong?

Answers will vary. For example: 7/12 is close to one-half. We subtract 1/4 from it — and the answer says we get 1/2! That does not make sense.

Add and Subtract: More Practice, cont.

5.

| $\frac{23}{36}$ | $\frac{5}{6}$ | $\frac{7}{15}$ | $\frac{3}{10}$ | $\frac{1}{10}$ | $\frac{4}{15}$ | $\frac{5}{6}$ | | $\frac{23}{30}$ | $\frac{7}{45}$ | | $\frac{8}{15}$ | $\frac{31}{30}$ | $\frac{17}{28}$ |
| B | E | C | A | U | S | E | | I | T | | W | A | S |

| $\frac{17}{21}$ | $\frac{9}{10}$ | $\frac{9}{35}$ | | $\frac{83}{72}$ | $\frac{11}{24}$ | $\frac{11}{24}$ | $\frac{27}{40}$ | $\frac{23}{30}$ | $\frac{5}{9}$ | $\frac{1}{6}$ | | $\frac{1}{14}$ | $\frac{104}{63}$ | $\frac{7}{6}$ | $\frac{13}{20}$ |
| N | O | T | | P | E | E | L | I | N | G | | W | E | L | L | ! |

Puzzle Corner

Grid 1:
$$\frac{2}{3} + \frac{1}{5} = \frac{13}{15}$$
$$+ \qquad +$$
$$\frac{1}{6} + \frac{1}{4} = \frac{5}{12}$$
$$= \qquad =$$
$$\frac{5}{6} \qquad \frac{9}{20}$$

Grid 2:
$$\frac{1}{6} + \frac{1}{7} = \frac{13}{42}$$
$$+ \qquad +$$
$$\frac{1}{8} + \frac{1}{9} = \frac{17}{72}$$
$$= \qquad =$$
$$\frac{7}{24} \qquad \frac{16}{63}$$

Adding and Subtracting Mixed Numbers, pp. 34-36

1.

a.
$6\frac{2}{3} \Rightarrow 6\frac{10}{15}$
$+\ 3\frac{1}{5} \qquad +\ 3\frac{3}{15}$
$\rule{2cm}{0.4pt} \qquad 9\frac{13}{15}$

b.
$10\frac{1}{8} \Rightarrow 10\frac{5}{40}$
$+\ 3\frac{2}{5} \qquad +\ 3\frac{16}{40}$
$\rule{2cm}{0.4pt} \qquad 13\frac{21}{40}$

c.
$17\frac{1}{16} \Rightarrow 17\frac{1}{16}$
$+\ 3\frac{3}{8} \qquad +\ 3\frac{6}{16}$
$\rule{2cm}{0.4pt} \qquad 20\frac{7}{16}$

2.

a.
$4\frac{1}{2} \Rightarrow 4\frac{5}{10}$
$+\ 3\frac{4}{5} \qquad +\ 3\frac{8}{10}$
$\rule{2cm}{0.4pt} \qquad 7\frac{13}{10} \Rightarrow 8\frac{3}{10}$

b.
$5\frac{5}{6} \Rightarrow 5\frac{5}{6}$
$+\ 7\frac{2}{3} \qquad +\ 7\frac{4}{6}$
$\rule{2cm}{0.4pt} \qquad 12\frac{9}{6} \Rightarrow 13\frac{3}{6}$

c.
$3\frac{5}{6} \Rightarrow 3\frac{20}{24}$
$+\ 2\frac{7}{8} \qquad +\ 2\frac{21}{24}$
$\rule{2cm}{0.4pt} \qquad 5\frac{41}{24} \Rightarrow 6\frac{17}{24}$

d.
$9\frac{5}{7} \Rightarrow 9\frac{15}{21}$
$+\ 7\frac{2}{3} \qquad +\ 7\frac{14}{21}$
$\rule{2cm}{0.4pt} \qquad 16\frac{29}{21} \Rightarrow 17\frac{8}{21}$

3.

a. $2\frac{6}{8} - 1\frac{3}{8} = 1\frac{3}{8}$

b. $3\frac{3}{6} - 1\frac{2}{6} = 2\frac{1}{6}$

c. $3\frac{3}{9} - 1\frac{4}{9} = 1\frac{8}{9}$

3

3

4.

a. $5\frac{1}{2} \Rightarrow 4\frac{15}{10}$ $-\ 2\frac{4}{5} \quad -\ 2\frac{8}{10}$ $\quad\quad\quad\quad 2\frac{7}{10}$	b. $15\frac{4}{8} \Rightarrow 14\frac{36}{24}$ $-\ 8\frac{5}{6} \quad -\ 8\frac{20}{24}$ $\quad\quad\quad\quad 6\frac{16}{24}$	c. $16\frac{5}{9} \Rightarrow 16\frac{10}{18}$ $-\ 10\frac{1}{2} \quad -\ 10\frac{9}{18}$ $\quad\quad\quad\quad 6\frac{1}{18}$
d. $4\frac{1}{6} \Rightarrow 3\frac{35}{30}$ $-\ 2\frac{3}{5} \quad -\ 2\frac{18}{30}$ $\quad\quad\quad\quad 1\frac{17}{30}$	e. $11\frac{1}{12} \Rightarrow 10\frac{13}{12}$ $-\ 3\frac{1}{4} \quad -\ 3\frac{3}{12}$ $\quad\quad\quad\quad 7\frac{10}{12}$	f. $8\frac{2}{9} \Rightarrow 7\frac{44}{36}$ $-\ 2\frac{3}{4} \quad -\ 2\frac{27}{36}$ $\quad\quad\quad\quad 5\frac{17}{36}$

5. a. Unreasonable answer. The sum should be more than 1/2, but 2/6 is less than half.
In reality, 1/2 kg + 1/4 kg = 3/4 kg
   b. Unreasonable answer. You cannot do more than 1 of a job. Also, 1/2 + 1/5 is less than 1 because 1/5 is quite a bit less than 1/2. In reality, 1/5 + 1/2 = 7/10.
   c. Reasonable, and correct.
   d. Reasonable, and correct.

6. a. 1 1/4 m + 8/10 m = 1 5/20 m + 16/20 m = 2 1/20 m
   b. 1.25 m + 0.8 m = 2.05 m.
   You will probably find that decimals are easier to work with.

7. a. 1 3/4 lb + 1 11/16 lb = 1 12/16 lb + 1 11/16 lb = 3 7/16 lb
   b. 3 7/16 lb = 3 lb 7 oz

Comparing Fractions, pp. 37-41

1. Answers will vary; check the student's answers, and give guidance while the student is working. For example:
   a. 5/12 is greater because it is more pieces than 3/12 (and they are the same kind of pieces).
   b. 4/5 is greater. It is the same number of pieces as 4/7 but each fifth is bigger than a seventh.
   c. 6/11 is greater because 1/2 is equal to 6/12, and between 6/12 and 6/11, 6/11 is greater.
   Or, 1/2 is equal to 5.5/11, so 6/11 is greater.
   d. 7/6 is more than 1, and 8/9 is less than one. So, 7/6 is more.
   e. 11/20 is more than 1/2, and 7/15 is less than 1/2, so the former is greater.
   f. Here, the student may not come up with a strategy, but one of the best ones is to convert both so they have the same denominator: 13/15 and 7/8 can be converted to 104/120 and 105/120. The latter is greater.

2. These fractions are more than 1/2: 5/8, 5/6, 8/14, 28/50.

3. a. >   b. <   c. >   d. <
   e. >   f. >   g. <   h. >
   i. <   j. <   k. =   l. <

4. a. <   b. >   c. <   d. <
   e. >   f. >   g. >   h. =

5. a. 16/24 > 15/24   b. 20/24 < 21/24   c. 10/30 > 9/30   d. 40/60 < 42/60
   e. 15/24 > 14/24   f. 55/40 < 56/40   g. 60/100 > 58/100   h. 54/45 < 55/45
   i. 49/70 < 50/70   j. 43/100 > 30/100   k. 63/56 < 64/56   l. 21/30 > 20/30

6. 1 1/2 cups is more than 1 1/3 cups so a triple batch of the first recipe uses more sugar (1/6 cup more).

Comparing Fractions, cont.

7. a. >   b. >   c. >   d. <
   e. <   f. >   g. <   h. <
   i. >   j. >   k. <   l. >

8. Since 1/4 < 3/10, taking off 3/10 of the price is bigger. The answer does not change if the price changes: 3/10 off of any price is a greater discount than 1/4 off of the same price.

9. a. $\dfrac{1}{5} < \dfrac{1}{3} < \dfrac{2}{5} < \dfrac{1}{2} < \dfrac{2}{3}$ b. $\dfrac{2}{2} < \dfrac{6}{5} < \dfrac{4}{3} < \dfrac{7}{5} < \dfrac{3}{2}$

10.

a. $\dfrac{7}{9} < \dfrac{7}{8} < \dfrac{9}{10}$	b. $\dfrac{2}{9} < \dfrac{1}{3} < \dfrac{4}{10}$

11. a. The women who swam were the bigger group. One-hundredth of 600 is 6; therefore 22/100 of 600 is 22 × 6 = 132. One-fifth of 600 is 120.

   b. Since 1/3 of them never exercise, 2/3 of them do exercise.

   c. Four hundred women of that group exercise.

   d. There are 132 women who swim. (See the answer for (a).)

Puzzle corner. If you cut the 12th parts into two, you can see that 1½ of the 12th parts is the same as 3/24, which is equal to 1/8. So either way, the dog would get the same amount.

Word Problems, pp. 42-43

1. a. Cindy makes two cakes, so she will need 7 dl of flour for those. In total she will need 7 dl + 5 dl + 3/4 dl = <u>12 3/4 dl</u>.
   b. Yes, 1 kg (which is 15 dl) of flour is enough.

2. The margins are a total of 3/4 in + 3/4 in = 1 1/2 inches. Subtract that from the width and height of the notebook. The width of the picture will be 3 1/4 in − 1 1/2 in = 1 3/4 in and the height will be 6 1/8 in − 1 1/2 in = 4 5/8 in.

3. 1 − 1/10 − 1/2 = 1 − 1/10 − 5/10 = 4/10. Jerry would do 4/10 of the project.

4. a. 19/100 + 2/10 = 19/100 + 20/100 = 39/100. This means he has 1 − 39/100 = 61/100 of his money left.
   b. 1/100 of $4,000 is $40. So, 61/100 of it is 61 × $40 = $2,440.

5. a. 1/8 + 1/8 + 1/4 + 1/4 = 1/8 + 1/8 + 2/8 + 2/8 = 6/8. Or, you can note that 1/8 + 1/8 makes 1/4, so in total they drank 3/4 of the smoothie. So, 2/8 or 1/4 of it is left.
   b. There is 500 ml (half a liter) of smoothie left. To find 2/8 of 2000 ml, you can first find 1/8 of it, which is 250 ml. Then, 2/8 of it is double that, or 500 ml.

Measuring in Inches, pp. 44-46

1.
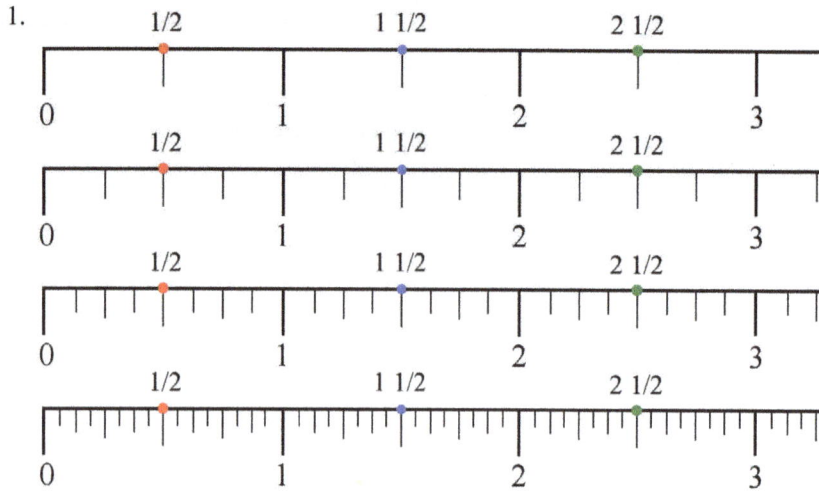

2.
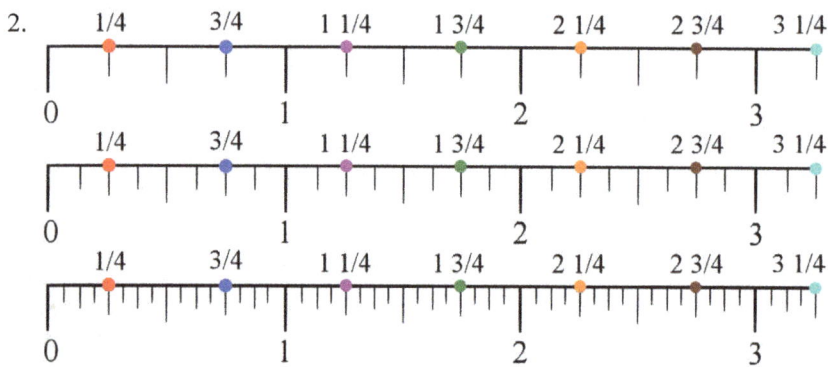

3.
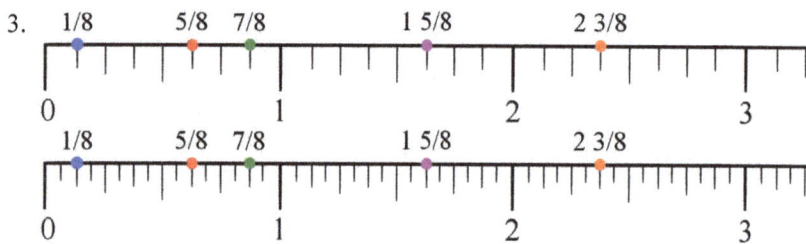

4.
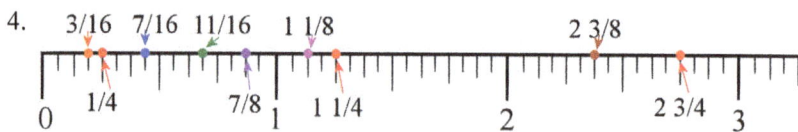

5. a. 1 1/4 in   b. 3/4 in
   c. 1 2/8 in or 1 3/8 in   d. 5/8 in or 6/8 in
   e. 1 5/16 in   f. 11/16 in
   g. 2 3/4 in
   h. 2 5/8 in
   i. 2 11/16 in

6. Answers will vary. Please check the students' work.

Measuring in Inches, cont.

7.

a. Using the 1/4-inch ruler: _3 or 3 1/4_ in Using the 1/8-inch ruler: _3 1/8_ in Using the 1/16-inch ruler: _3 2/16 = 3 1/8_ in	b. Using the 1/4-inch ruler: _3/4_ in Using the 1/8-inch ruler: _5/8 or 6/8_ in Using the 1/16-inch ruler: _11/16_ in
c. Using the 1/4-inch ruler: _2 3/4_ in Using the 1/8-inch ruler: _2 6/8 = 2 3/4_ in Using the 1/16-inch ruler: _2 12/16 = 2 3/4_ in	d. Using the 1/4-inch ruler: _1_ in Using the 1/8-inch ruler: _1_ in Using the 1/16-inch ruler: _15/16_ in
e. Using the 1/4-inch ruler: _1 1/4_ in Using the 1/8-inch ruler: _1 3/8_ in Using the 1/16-inch ruler: _1 5/16_ in	f. Using the 1/4-inch ruler: _1 3/4_ in Using the 1/8-inch ruler: _1 6/8 = 1 3/4_ in Using the 1/16-inch ruler: _1 12/16 = 1 3/4_ in

Line Plots and More Measuring, pp. 47-49

1. a. Three days.
   b. 1 hour 50 minutes
   c. 9 hours 20 minutes

2. a.

   b. ¾ C + ¾ C + ¾ C = 2 ¼ C
   c. 1 ¾ C + 1 ¾ C + 1 ¾ C = 5 ¼ C

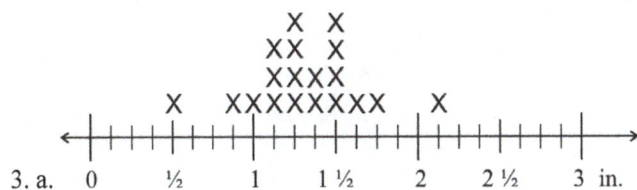

3. a.

   b. Those that are strictly less than 1 ½ inches long.

   c. 2 ⅛ in + 1 ¾ in + 1 ⅝ in = 2 ⅛ in + 1 6/8 in + 1 ⅝ in = 4 12/8 in = 5 4/8 in or 5 ½ in.

4. a. The sides measure: 1 ½ in, 3 15/16 in, 1 11/16 in, and 3 ¼ in.
   b. The perimeter is 10 ⅜ in.

5. The easiest way is to check how much longer each side is. The 3 ⅛ side is ⅛ in. longer than the 3-inch side. The 6 ¼-side is ⅛ inch longer than the 6 ⅛-inch side.
   There are four sides. The perimeter of the first phone is ⅛ + ⅛ + ⅛ + ⅛ = 4/8 = ½ inch longer than the perimeter of the second.

6. Answers will vary. Please check the students' work.

Review, pp. 50-52

1. a. 19/2   b. 61/11   c. 58/7   d. 506/100

2. a. 4 1/10   b. 6 1/3   c. 3 1/9   d. 2 8/12

3. 23/6 = 3 5/6

4. a. 5 5/8   b. 6 12/20   c. 5 4/15

5. a. 15/21 + 7/21 = 22/21 = 1 1/21   b. 9/30 + 10/30 = 19/30
   c. 2 9/7 − 1 6/7 = 1 3/7   d. 2 16/20 + 3 5/20 = 5 21/20 = 6 1/20

6. a. <   b. <   c. =   d. <   e. <   f. >   g. <   h. <

7. From the first piece, she has left: 5 1/2 ft − 3 1/8 ft = 5 4/8 ft − 3 1/8 ft = 2 3/8 ft.
   From the second piece, she has left: 4 1/4 ft − 3 1/8 ft = 4 2/8 ft − 3 1/8 ft = 1 1/8 ft.
   Combined, those two pieces are 2 3/8 ft + 1 1/8 ft = 3 4/8 ft = 3 1/2 ft.

8. 1 − 32/100 − 42/100 − 2/10 = 1 − 32/100 − 42/100 − 20/100 = 6/100. So, 6/100 of the land is resting.

9. One-fifth of $35 is $7, so the discounted price would be $28. And 2/11 of $33 is $6, so its discounted price would be $27. So, 2/11 off of the $33-book is the better deal.

   If both books cost $50, then 1/5 off of it would be the better buy. This is because 1/5 is more than 2/11.

10. a. two beakers
    b. one beaker
    c. 1 7/8 + 2 + 2 + 2 + 2 1/8 = 10 cups

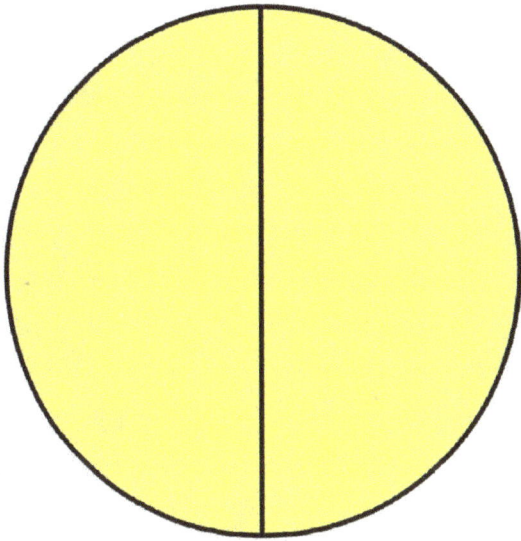
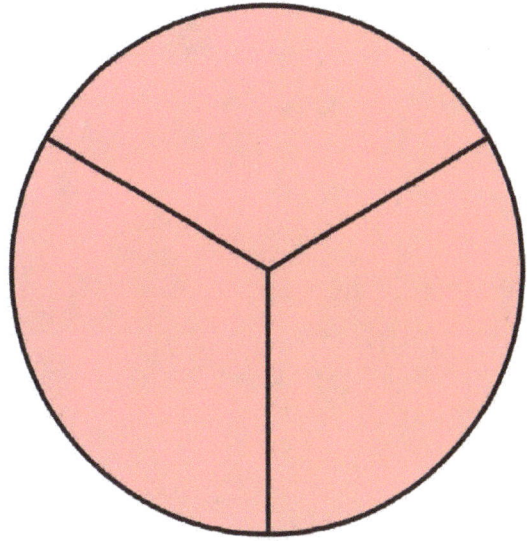
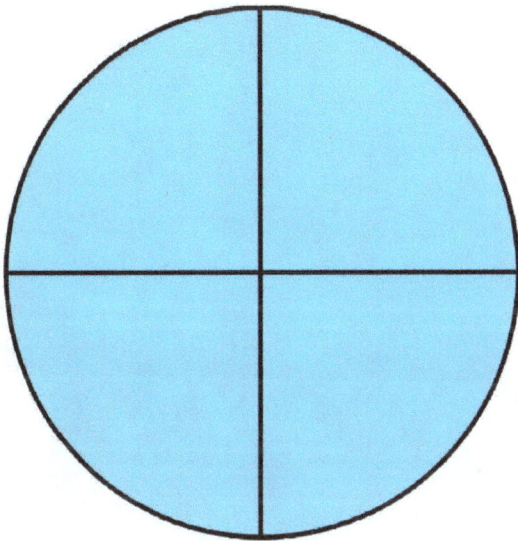
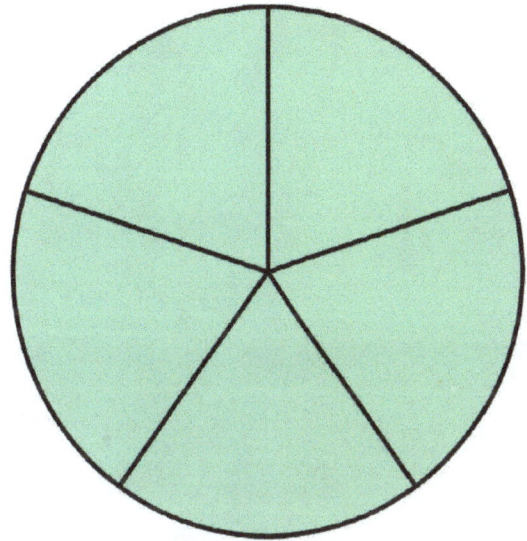
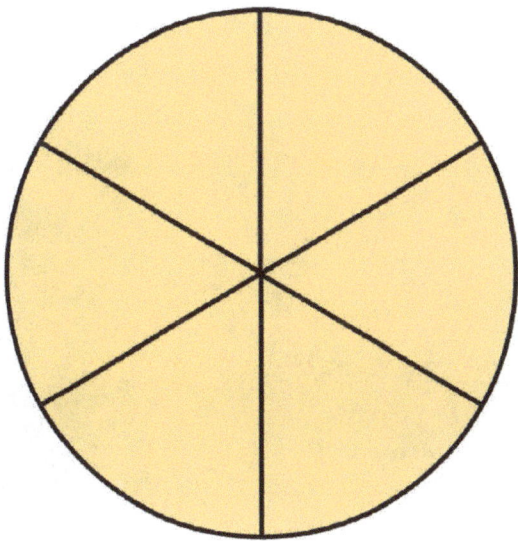
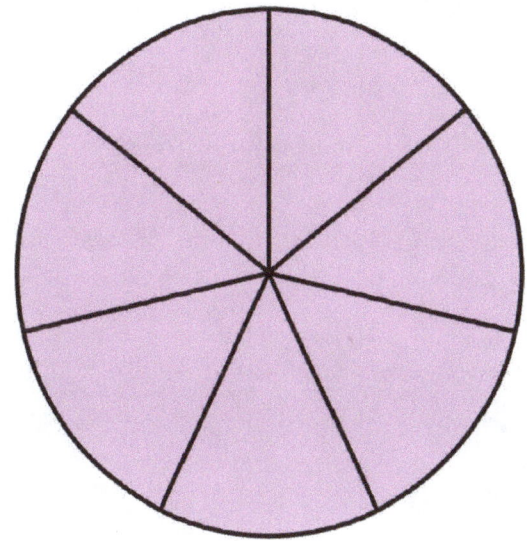

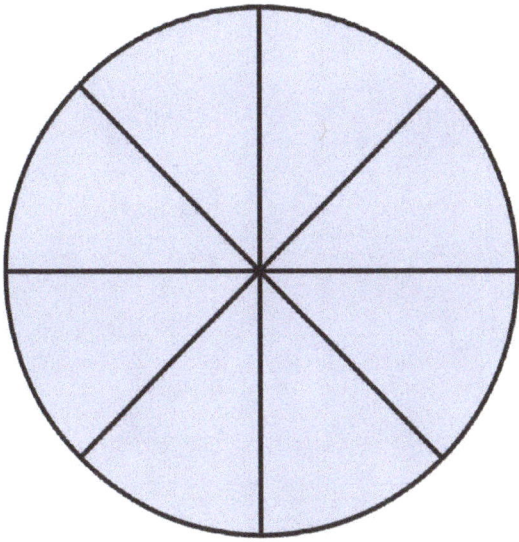

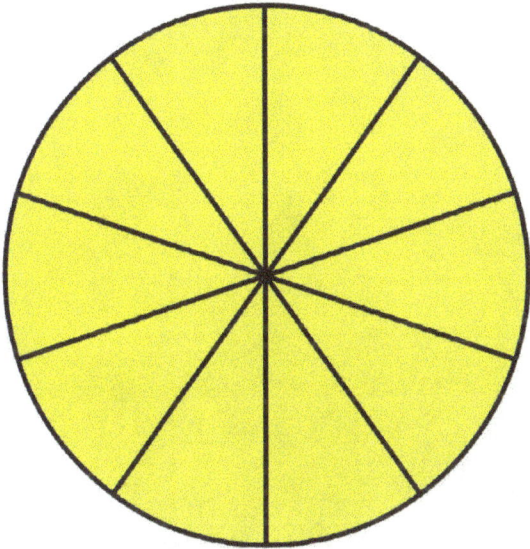
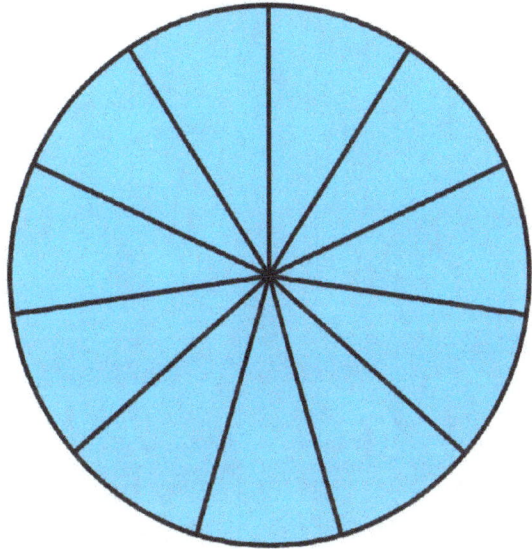
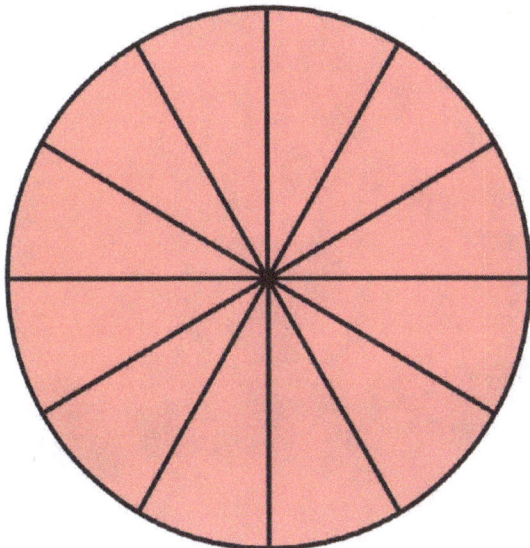
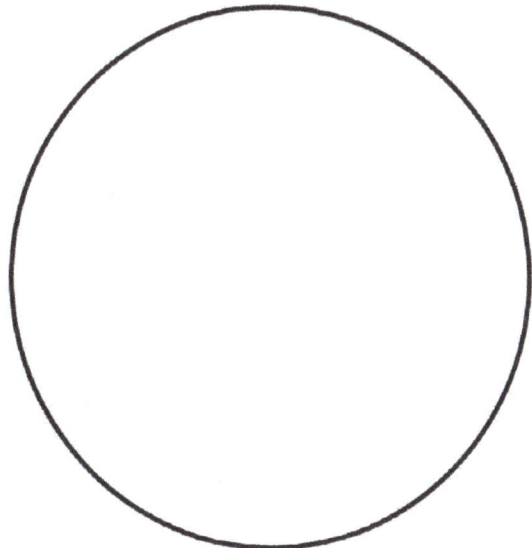

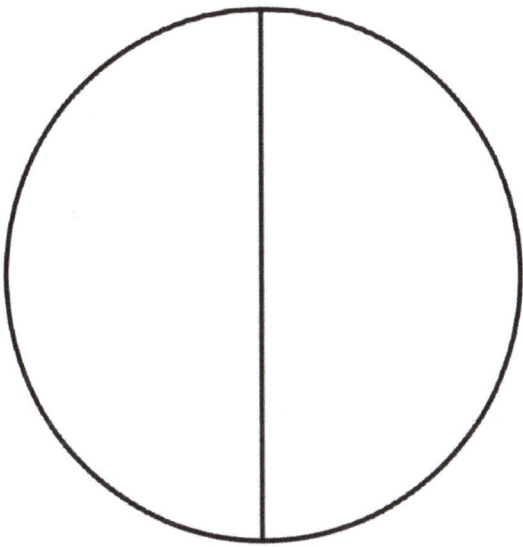 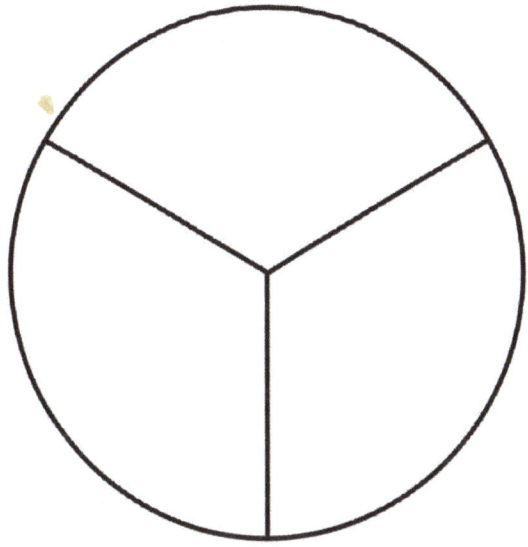

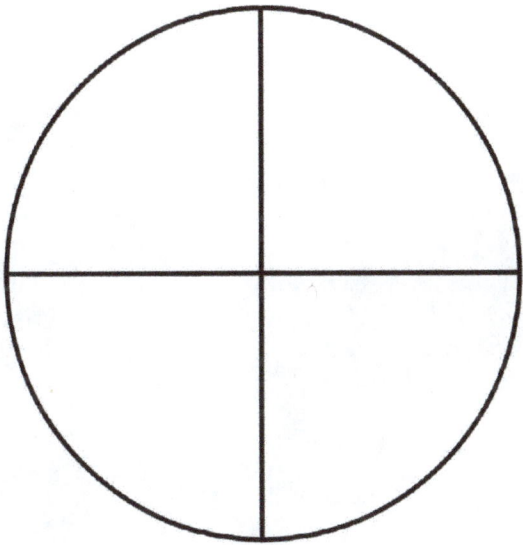 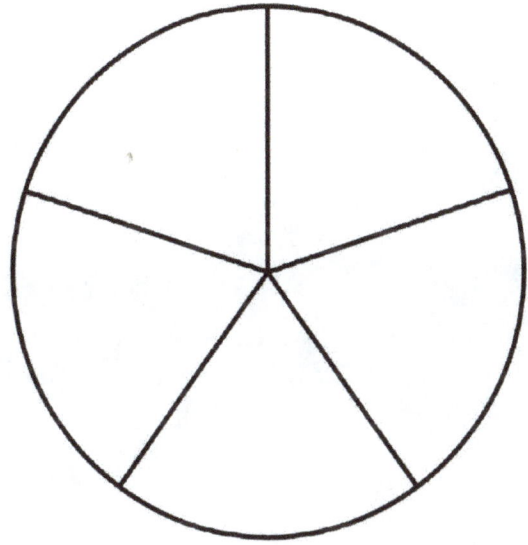

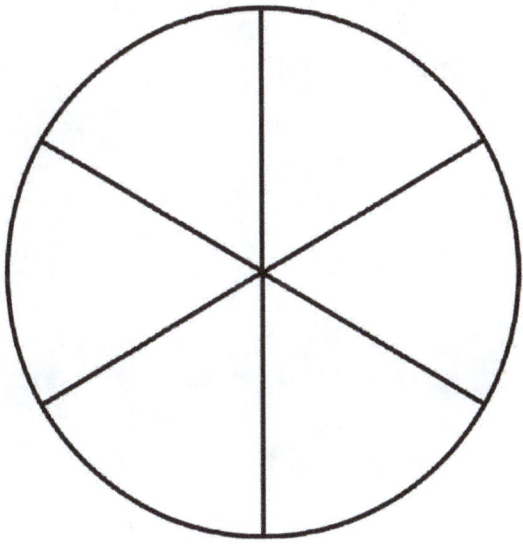 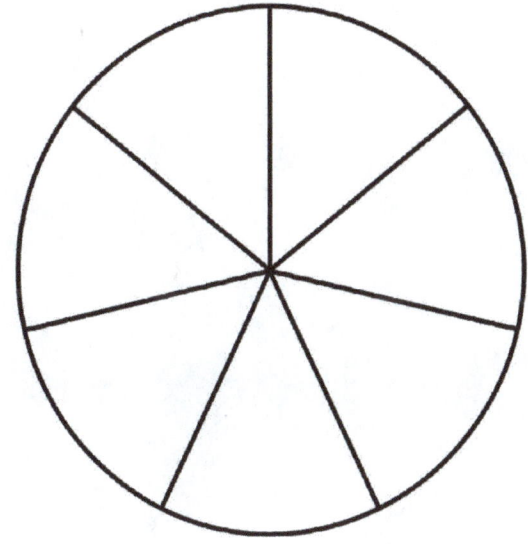

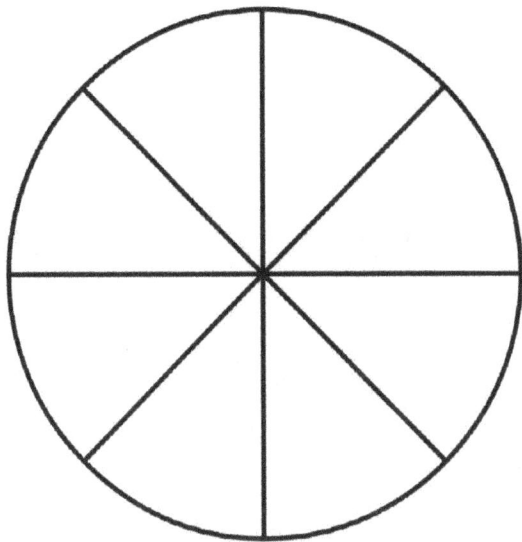
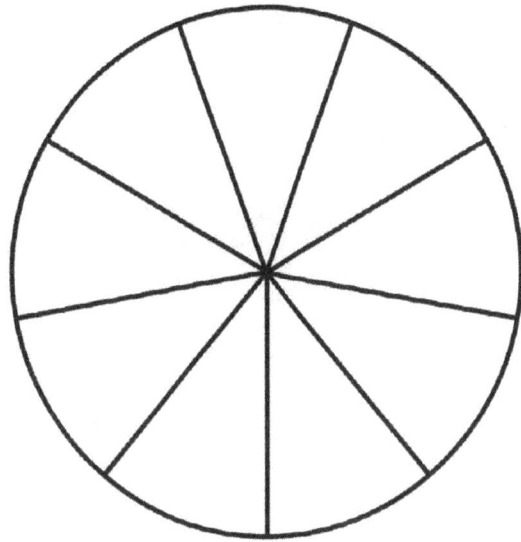
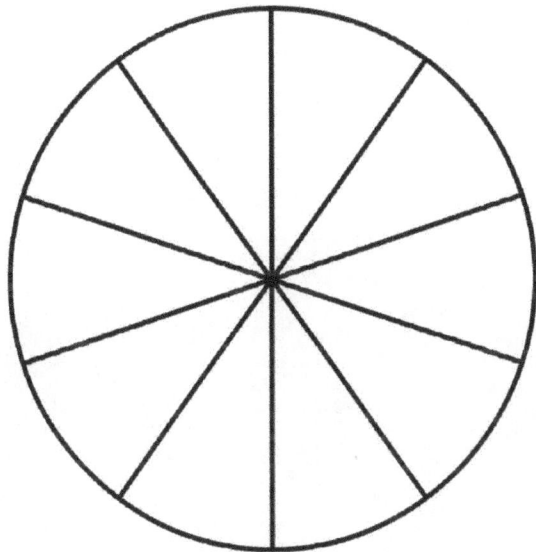
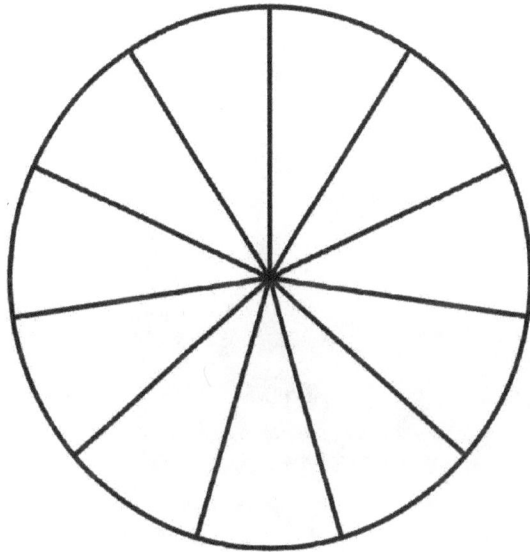
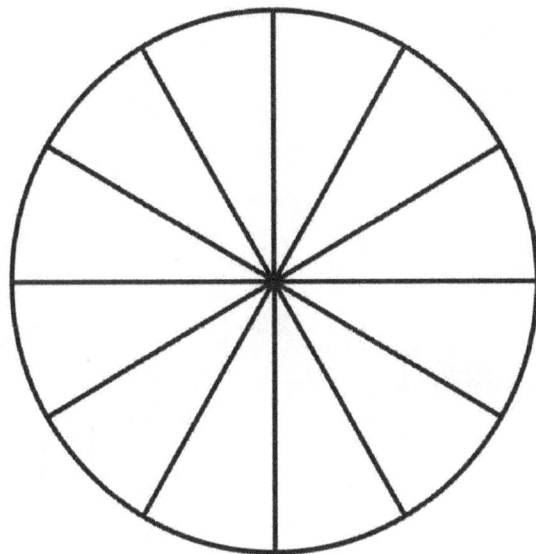
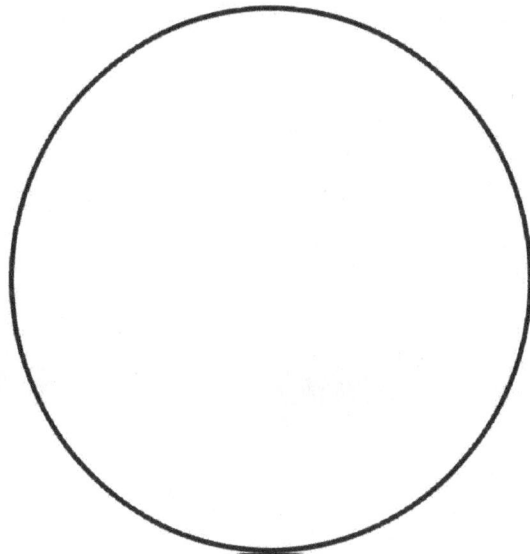

More from math MAMMOTH

Math Mammoth has a variety of resources to fit your needs. All are available as economical downloads, and most also as printed copies.

- **Math Mammoth Light Blue Series**
  A complete curriculum for grades 1-7. Each grade level includes two student worktexts (A and B), which contain all the instruction and exercises all in the same book, answer keys, tests, cumulative reviews, and a worksheet maker. International (all metric), Canadian, and South African versions are also available.
  https://www.MathMammoth.com/complete-curriculum

  https://www.MathMammoth.com/international/international

  https://www.MathMammoth.com/canada/

  https://www.MathMammoth.com/south_africa/

- **Math Mammoth Skills Review Workbooks**
  These workbooks are intended to be used alongside the Light Blue series full curriculum, and they provide additional review to the topics studied in the main curriculum, in a spiral manner.
  https://www.MathMammoth.com/skills_review_workbooks/

- **Math Mammoth Blue Series**
  Blue Series books are topical worktexts for grades 1-7, containing both instruction and exercises. The topics cover all elementary mathematics from 1st through 7th grade. These books are not tied to grade levels, and are thus great for filling in gaps.
  https://www.MathMammoth.com/blue-series

- **Make It Real Learning**
  These activity workbooks concentrate on answering the question, "Where is math used in real life?" The series includes various workbooks for grades 3-12.
  https://www.MathMammoth.com/worksheets/mirl/

- **Review Workbooks**
  Workbooks for grades 1-7 that provide a comprehensive review of one grade level of math—for example, for review during school break or summer vacation.
  https://www.MathMammoth.com/review_workbooks/

Free gift!

- Receive over 350 free sample pages and worksheets from my books, plus other freebies:
  https://www.MathMammoth.com/worksheets/free

Lastly...

- Inspire4 is an inspirational website for the whole family I've been privileged to help with:
  https://www.inspire4.com

www.ingramcontent.com/pod-product-compliance
Lightning Source LLC
Chambersburg PA
CBHW081237090426
42738CB00016B/3339